GW01385438

Oxford

PROBABILITY & STATISTICS

1

for Cambridge International AS & A Level

Series editor: David Rayner

James Nicholson

Oxford and Cambridge
leading education together

OXFORD
UNIVERSITY PRESS

OXFORD
UNIVERSITY PRESS

Great Clarendon Street, Oxford, OX2 6DP, United Kingdom

Oxford University Press is a department of the University of Oxford. It furthers
the University's objective of excellence in research, scholarship, and education
by publishing worldwide. Oxford is a registered trade mark of Oxford University
Press in the UK and in certain other countries

British Library Cataloguing in Publication Data

Data available

978-0-19-830693-1

10 9 8 7 6 5 4 3 2 1

MIX
Paper from
responsible sources
FSC
www.fsc.org
FSC® C007785

Paper used in the production of this book is a natural, recyclable product made from wood
grown in sustainable forests. The manufacturing process conforms to the environmental
regulations of the country of origin.

Printed by Bell and Bain Ltd, Glasgow

The publisher would like to thank Cambridge International Examinations for their
kind permission to reproduce past paper questions.

*Cambridge International Examinations bears no responsibility for the example answers to
questions taken from its past papers which are contained in this publication.*

*The questions, example answers, marks awarded and comments that appear in this book
were written by the authors. In examination, the way marks would be awarded to
answers like these may be different.*

Acknowledgements

The publishers would like to thank the following for permissions to use their photographs:

p2: nevio/Shutterstock; **p3:** Myibean/Shutterstock; **p4t:** joeysworld.com/Alamy; **p4b:** Milos
Muller/Shutterstock; **p5tr:** Ricky John Molly/The Image Bank/Getty; **p5cl:** imageBROKER/Alamy;
p5c: Russel Shively/Shutterstock; **p5cr:** Africa Studio/Shutterstock; **p5b:** JamesChen/Shutterstock;
p7: AFP/Getty; **p8t:** Sally and Richard Greenhill/Alamy; **p8b:** Brandon Bourdages/Shutterstock;
p9: Alex Coombs/OUP; **p10:** David Spurdens/extremesportsphoto.com/Corbis; **p11:** Giancarlo
Liguori/Shutterstock; **p12tl:** Justin Kase zsixz/Alamy; **p12tr:** PHOTOMDP/Shutterstock; **p12cl:**
terekhov igor/Shutterstock; **p12cr:** ChameleonsEye/Shutterstock; **p12bl:** Air Images/Shutterstock;
p12br: RGB Ventures/Superstock/Alamy; **p13tr:** elen_studio/Shutterstock; **p13tl:** Domenic
Gareri/Shutterstock; **p13cl:** Greg and Jan Ritchie/Shutterstock; **p13cr:** Vladimir Korostyshevskiy/
Shutterstock; **p13bl:** leungchopan/Shutterstock; **p13br:** Gritsana P/Shutterstock; **p14:** James
Boardman/Alamy; **p25:** Richard Sustano/Shutterstock; **p68:** gapminder.org; **p70:** Bildagentur
Zoonar GmbH/Shutterstock; **p92:** testing/Shutterstock; **p108:** Tim Gartside/Alamy; **p132t:** Michael
Steele/Getty; **p132b:** Michael Steele/Getty; **p134:** focuslight/Shutterstock; **p152:** Trevor Reeves/
Shutterstock; **p153t:** Gay Pride Parade/Alamy; **p153b:** mady70/Shutterstock; **p155l:** ImagineChina/REX;
p155r: New York Daily News Archive/Getty; **p155b:** Clive Brunskill/Alamy; **p169:** Brian Jackson/Alamy;
p178: MBI/Alamy; **p191t:** Anton Balazh/Shutterstock; **p191b:** Peter Macdiarmid/Getty

Contents

Introduction

About this book

This book has been written to cover the **Cambridge AS and A Level International Mathematics (9709)** course, and is fully aligned to the syllabus.

In addition to the main curriculum content, you will find:

- 'Maths in real-life', showing how principles learned in this course are used in the real world.
- Chapter openers, which outline how each topic in the Cambridge 9709 syllabus is used in real-life.

The book contains the following features:

Notes	Did you know?
Advice on calculator use	EXAM STYLE QUESTION

Throughout the book, you will encounter worked examples and a host of rigorous exercises. The examples show you the important techniques required to tackle questions. The exercises are carefully graded, starting from a basic level and going up to exam standard, allowing you plenty of opportunities to practise your skills. Together, the examples and exercises put maths in a real-world context, with a truly international focus.

At the start of each chapter, you will see a list of objectives that are covered in the chapter. These objectives are drawn from the Cambridge AS and A Level syllabus. Each chapter begins with a *Before you start* section and finishes with a *Summary exercise* and *Chapter summary*, ensuring that you fully understand each topic.

Review exercises are placed after chapters 3, 6 and 9. They comprise a host of exam questions which cover the topics from the previous 3 chapters and are in no particular order of difficulty.

The answers given at the back of the book are concise. However, when answering exam-style questions, you should show as many steps in your working as possible. All exam-style questions, as well as *Paper A* and *Paper B*, have been written by the author.

About the authors

James Nicholson is an experienced teacher of mathematics at secondary level; he has having taught for 12 years at Harrow School and spent 13 years as Head of Mathematics in a large Belfast grammar school. He is the author of two A-level statistics texts, and editor of the *Concise Oxford Dictionary of Mathematics*. He has also contributed to a number of other sets of curriculum and assessment materials, is an experienced examiner and has acted as a consultant for UK government agencies on accreditation of new specifications.

James ran school workshops for the Royal Statistical Society for many years, and has been a member of the Schools and Further Education Committee of the Institute of Mathematics and its Application since 2000, including 6 years as chair; and is currently a member of the Outer Circle group for the Advisory Committee on Mathematics Education. He has served as a Vice-President of the International Association for Statistics Education for 4 years, and is currently Chair of the Advisory Board to the International Statistical Literacy Project.

A note from the author

The aim of this book is to help students prepare for the Statistics 1 unit of the Cambridge International AS and A Level Mathematics syllabus, although it may also be useful in providing support material for other AS and A Level courses. The book contains a large number of practice questions, many of which are exam-style in addition to questions from past Cambridge examinations papers.

In writing the book I have drawn on my experiences of teaching Mathematics, Statistics and Further Mathematics to A Level over many years, as well as on my experience as an examiner and discussions with statistics educators from many countries at international conferences.

Statistical data have been recorded for most of the time civilization has existed. The Romans gathered a lot of information on the population size and wealth of nations in their empire. They usually carried out a census every five years, which required citizens to register their duties and property; this was then used to calculate taxes. Modern statistics, which attempts to do more than just record information, dates back to the seventeenth century.

Objectives

- Describe the role of statistics in modern society.
- Identify types of data: primary and secondary, categorical and numerical, discrete and continuous.

1.1 What is statistics and why is it important?

We live most of our lives relying on only partial information to make decisions. Some of the information we use is numerical, other information can be summarized numerically.

As we don't often have access to all the information when making decisions, statistics is largely about trying to make the best use of the data we have at any given time. If we have a better understanding of the information we have available, this generally helps us to make better decisions because we have a better understanding of the risks involved.

Consider the graph on the right, which shows the body temperature readings for 20 people. Each person has had their temperature taken with a mouth ('oral') and an ear ('tympanic') thermometer.

Notice that for each person there are variations between the oral and tympanic readings, even though they claim to be testing the same value (body temperature).

Body temperature measurements

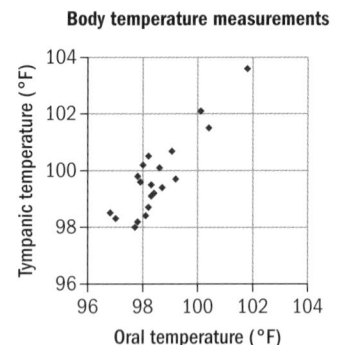

The next graph shows the temperatures of the same 20 people but now the gender of each person is identified. We see that there is even more variability: it appears that males and females may have different body temperature characteristics.

We are surrounded by variation – essentially the unpredictability of life. Statistics is a tool for helping us to make sense of the partial information we have available and make more informed decisions. We already do this naturally and informally in many everyday situations.

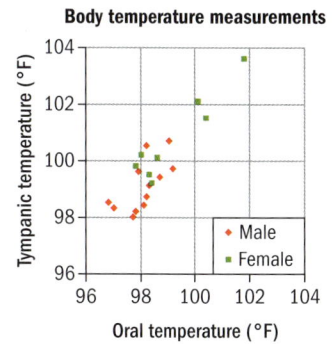

Body temperature measurements

Consider the following examples involving uncertainty:

Example 1

The insurance industry is based on a business contract that is beneficial to both the individual and the company. Why do we have insurance?

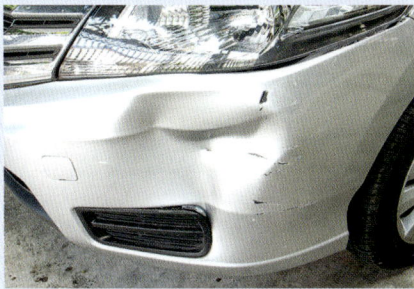

- It is good for us: a big loss may be unlikely to happen, but we couldn't afford it if it did.
- It is good for the insurance companies: the large number of premium they take in pay for the small number of claims made, which means they will make a profit.

Although insurance involves uncertainty for both parties, statistics can be used to calculate the potential risk and rewards involved.

Example 2

A lift should have a safety notice giving the number of people and the weight it is able to accommodate safely. How should the company who makes the lift calculate these?

The number of people allowed in the lift at one time is only a **proxy** for the important variable – the total weight. The company who makes the lift will have to assume an average weight for each person; however, their weight could vary dramatically. The reality is that all mathematics of combining distributions depends on assumptions of independence, and we must keep the presence of variation in mind.

Exercise 1.1

Consider the following situations, where there may be a considerable degree of variation.

1. Why is a knock-out competition much more likely to produce a 'surprise' winner than an extended league format?

Knock-out competition

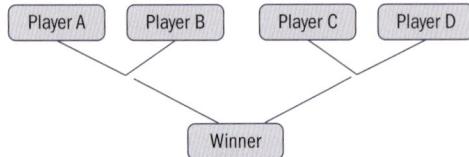

League	Player A	Player B	Player C	Player D
Player A				
Player B				
Player C				
Player D				

The knock-out competition may produce a winner, but does this mean that they are statistically the best player?

2. A haulage firm has to quote a price for a job in which one of the major costs is the length of time the journey will take. They have to quote in advance, sometimes quite a long time in advance, but don't know how long the journey will take until it happens. What advice would you give to them?

3. The government are to set up an enquiry into health and safety issues relating to the positioning and operation of mobile phone transmitters. What factors should they be looking at?

Apart from the presence of variation, we need to be conscious that different contexts can tolerate different amounts of variation. For example:

A newspaper does not need the same precision in printing as a banknote.

There can be more variation in the manufacture of generic travel sickness tablets than of powerful chemotherapy drugs.

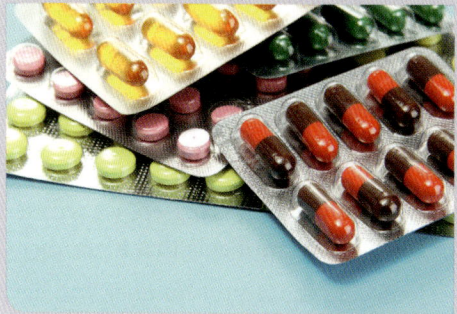

Exercise 1.2

We know from experience that when a die is thrown repeatedly, we will see different scores, but we usually don't think about this systematically. It is often difficult to judge how the scores of a die will vary, as we will see in the following exercise.

Tragic consequences may follow if a change in the quality of steel girders produced for use in bridges or large buildings is not immediately picked up and rectified.

Carry out the following experiments, and then consider the questions:

A Toss a fair coin 10 times and count the number of heads that are seen. Repeat this at least 5 times.

B Throw a fair die 20 times and record the results using a tally chart. Make a table showing the frequencies of each score: 1, 2, 3, 4, 5 and 6. Repeat this so you have at least 3 sets of frequencies.

C Toss a fair coin and count the number of tosses until a tail is seen. Record '1' if a tail is seen on the first toss, record '2' if a tail is seen on the second toss, and so on. Repeat this at least 10 times.

If you are working as a class, bringing your results together will give you a much larger data set to consider.
Keep your results, as you will need them in Chapter 2 to calculate averages and some formal measures of spread.

1. When a fair coin is tossed, we will on average see five heads. Did you get five heads most of the time for experiment A? What proportion of the time would you expect to get five heads?

2. When a fair die is thrown 20 times, we will not see all the possible outcomes an equal number of times. We will see each score 'on average' between three and four times. When you did Experiment B, did all the possible scores occur three or four times? Were there any scores which did not come up at all in a complete group of 20 throws?

   If a die is rolled 20 times, what proportion of the time should we expect to produce a zero frequency for a particular score? i.e. for a particular score to not be rolled.

3. When a fair coin is tossed, we will see a tail on the first toss about half the time, so it is probable that a number of 1s were seen in Experiment C.

   Consider the largest number you have in your list. If this experiment was repeated many times, how often should we expect to see a 4 appear in the list? Is it possible that 100 would appear?

Sampling

If you go on to study S2 or work with statistics later, you will encounter sampling theory. You have probably considered simple aspects of this already in your study of statistics – the need to avoid bias, for example. So why do we use samples?

- To save time and money. Often sampling will give us a large proportion of the total information for a fraction of the cost – certainly a law of diminishing returns applies to increasing the sample size.
- It is often not possible to have complete information, since

   some tests involve the destruction of the item

   testing may be very expensive

   there just may not be time.

The purpose for which the information is to be used influences how accurate we feel the information needs to be. Compare, for example, information which is needed

- to assist marketing strategy
- to monitor health and safety.

Consider the following situations:

- An airline has 243 passengers booked on the last flight of the day, from Sydney to Seoul.

   If the airline has to cancel the flight, how many of those passengers will want to travel to Seoul the next day?

   How many people will the airline have to provide overnight accommodation for?

It is not realistic to expect to know in advance exactly how many of the 243 passengers will want to travel the next day, or how many will want overnight accommodation. However, if the airline has some historical data on what has happened in similar circumstances on previous occasions, it can start making contingency plans. Many airlines now ask whether the passenger is travelling for business, to visit family or for leisure when they book tickets.

Probability distributions will help you understand what happens in these situations.

- A rare but very serious disease occurs in 1 in 10 000 people. There is a screening test which gives a positive result in 99 % of cases where the subject has the disease and gives a negative result in 99 % of cases where the subject does not have the disease. How reasonable is it to tell a patient whose test comes back positive that they have this serious disease?

You will look at screening tests again, in more detail, on pages 77 and 79

Even though this screening test is extremely accurate, both for those with the disease and those without, for every person whose positive result occurred because they had the disease, there will be around 100 people who gave a positive result without having the disease.

Conditional probability will help you understand what the implications of a positive result in a situation like this actually are.

- There is substantial evidence that shows that students in large classes in the UK have better examination results than students in small classes. Would it be reasonable for the UK government to decide that all students should be taught in large classes, in an effort to improve educational standards?

This probably would not be a good idea – schools put students who are more capable academically into larger classes, and their academic ability is the reason they get better results.

This is an example where simply observing what is going on, instead of conducting a designed **statistical experiment**, could be very misleading.

1.2 Types of data

> Data can either be **quantitative** (numerical) or **qualitative** (non-numerical). **Categorical data** is data which can be divided into distinct categories, and can be either qualitative or quantitative.

The fruit on this trader's stall can be described by its properties.

For example:

The type of fruit

The colour

How ripe it is

These are **qualitative data** – non-numerical or **categorical** data

The number of that type of fruit

The weight of a piece of fruit

The cost of the fruit

These are **quantitative data** – numerical

Quantitative data can either be discrete (limited to particular values) or continuous (any value within a range).

There are two types of quantitative data.

The number of pieces of fruit can only be one of a list of values (0, 1, 2, …). For example, these 5 oranges are an example of **discrete data**.

The weight of the oranges can be read from the scales as 1400 grams: If you measured it more exactly, however, you might find it is 1396.2 g (to 1 d.p.).
This is an example of **continuous data** – it will always be reported as a rounded value.

Example 3

What data about the radio are given in the advertisement?

For each piece of data, state whether it is qualitative or quantitative, and, if quantitative, whether it is discrete or continuous.

Puretone DAB/FM radio $ 89.99

Power output	2 × 2 W
Wavebands	DAB/FM
Dimensions	25 cm × 20 cm × 5 cm
Colour	Available in black or silver
Mains/battery	Mains, or 4 × AA batteries

▶ Continued on the next page

Power: quantitative, continuous
Wavebands: qualitative
Number of batteries needed: quantitative, discrete
Price: quantitative, discrete
Colour: qualitative
Length, width, height: quantitative, continuous

For a large number of data observations, a list is not easy to make sense of.

A **frequency table** of values can be used to summarize a large list.
Sometimes it is helpful to use groups of values so the data are more
easily understood.

Example 4

A group of 45 women go on a skiing
holiday.

Summarize the following data in frequency
tables.

a) The number of children each woman has:

1	0	2	1	2	0	0	3	1	1	2	1	0	0	1
2	1	1	0	1	1	1	0	2	5	1	2	1	0	1
2	1	3	1	1	0	0	1	2	1	1	3	1	0	0

b) The women's ages (use **class intervals** of 15–19, 20–24, ..., 45–49):

22	26	26	45	26	27	17	29	35	38	25	23	17	38	48
26	28	35	32	19	28	35	17	29	36	32	34	27	19	25
28	23	35	24	29	37	20	30	44	25	19	32	22	22	34

▶ Continued on the next page

a)

Number of children	Frequency
0	12
1	21
2	8
3	3
4	0
5	1

12 women have no children

b)

Age	Frequency
15–19	6
20–24	7
25–29	15
30–34	6
35–39	8
40–44	1
45–49	2

In this table, grouping the data sacrifices detail but allows you to get a sense of the distribution

Example 5

Consider the marina at Monte Carlo, list at least four pieces of relevant data that might be collected. Include the type of data, and how it may be collected.

Examples of data that could be collected for this situation include the number of boats in the marina at any time (quantitative, discrete) and, for each boat:

- the colour (qualitative)
- the length (quantitative, continuous)
- the age of the boat (quantitative, discrete)
- the name (qualitative)
- the value (quantitative, discrete – though very many values are possible).

The number of boats and colour could simply be observed. The length could be measured, though perhaps only approximately if you can't get permission to board the boat. The age of the boat, however, might need sight of a register not freely available and the value can probably only be estimated by making comparisons with other similar boats sold in the recent past – and Monte Carlo marina probably has some boats in it for which there are very few comparable boats, making the estimate difficult. Colour, name and length are potential primary data examples, while the year of commission and the value are likely to be available as secondary data only.

Primary data is any data which you collect yourself – it may be measurements but can also be through questionnaires or other surveys, investigations and experiments.

Secondary data has been collected by someone other than the person using the data – which may be in the form of the raw data, but it also includes databases, published statistics, newspapers etc.

The internet has a huge amount of secondary data freely available for people to access, for example National Statistics, but also data from organisations such as the Guinness Book of Records are available online. With secondary data it is important to know that it was collected reliably, and that the data matches what you want to investigate – if it does not have all the information you need, you will not be able to use it. However, when it is available secondary data saves a lot of time and effort. A lot of data is now collected automatically (**data logging**) using some form of technology.

> Any time you use data in a report you should specify the source – either where you located secondary data, or how you collected it yourself. This allows anyone reading your analysis of the data to know what sort of evidence it is based on.

Example 6

Amir is concerned about the traffic passing his house. One morning, he spends an hour counting the cars that pass his house. He then hears that the council have installed a sensor which automatically registers when a car goes past. Amir gets a summary of the traffic in the past six months from the council.

· ·

The first set of data described is primary because Amir collected it himself. The second set is secondary data because it was collected by someone else (the council) in a data logging process.

Exercise 1.3

1. The number of days each student in a class was late during one week was recorded. Summarize these data in a frequency table.

0	1	0	0	1	3	0	0	0	0
0	5	1	0	1	0	0	0	0	2
0	1	1	0	0	0	1	2	0	0

> Data from *NOAA Tides and Currents*, 30 March 2014. http://www.tidesandcurrents.noaa.gov (Note that there are only 29 high tides in 15 days because the time between successive high tides is something over 12 hours – the time of high tide on a particular day is around 45 minutes later than on the previous day.)

2. The depth of water (in feet) at high tide was recorded at San Francisco, Golden Gate for a period of 15 days in January 2012. Summarize the data, listed here, in a grouped frequency table. Use appropriate class intervals.

5.5	3.6	5.6	3.6	5.7	3.9	5.8	4.1	6.0	4.3
6.2	4.5	6.3	4.7	6.4	4.9	6.4	5.0	6.3	5.2
6.1	5.4	5.7	5.6	5.2	5.8	4.6	5.9	4.1	

3. For each of the following situations, list at least four pieces of relevant data that may be collected. Include at one type of primary data and one type of secondary data, stating what type it is and how it may be collected.

A class can work on this part of the exercise in groups, each group looking at a given number of questions. If working on your own, try the first three or four questions and then compare your answers with the answer section – the suggestions listed there are unlikely to be the same as yours, but you can make sure your list contains the same sorts of things.

a) Some cars for sale on a showroom forecourt.

b) A motorcycle road race.

c) The headlights of a car.

d) A river in New Zealand.

e) The wool from a flock of sheep.

f) Chocolate cakes produced in a Swiss factory.

g) The snow that fell in a snowstorm.

h) A baseball player.

i) A grizzly bear.

j) The Great Sphinx of Giza, in Egypt.

k) The Hong Kong business district.

l) Traffic in Bangkok.

Chapter summary

- Data can either be quantitative or qualitative.
- Quantitative data is numerical in kind.
- Qualitative data is non-numerical in kind.
- Quantitative data can either be discrete (limited to particular values) or continuous (any value within a range).
- A frequency table of values can be used to summarize a large number of data observations. Groups of values may be used so the data are more easily understood.

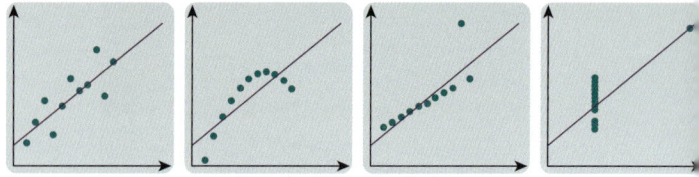

Frank Anscombe (1918–2001) was a British statistician who illustrated brilliantly why we should be careful not to rely too heavily on summary statistics. The four data sets shown here, despite being very different, have identical summary statistics – the mean and variance of x and y are the same, and they have the same line of best fit and correlation. Extreme values can get lost in summary statistics but often have a huge impact – a freak storm for instance.

Objectives

- Understand and use different measures of central tendency (mean, median and mode) and variation (range, interquartile range and standard deviation), for example when comparing and contrasting sets of data.
- Calculate the mean and standard deviation of a set of data (including grouped data), either from the set of data itself or from given totals such as $\sum x$ and $\sum x^2$, or $\sum (x - a)$ and $\sum (x - a)^2$.

Before you start

You should know how to:

1. Calculate the mean, median, range and mode of a simple set of data.
   For example: 7, 5, 3, 4, 7, 5, 7, 5, 9, 6, 5
   The median is the middle value with the data in order: 3, 4, 5, 5, 5, **5**, 6, 7, 7, 7, 9. Here the median is 5.
   Range = highest value – lowest value:
   Range = 9 – 3 = 6.
   The mode is the value which appears most often: Here the mode is 5.

   $$\text{Mean} = \frac{\text{sum of values}}{\text{number of values}}$$

   $$= \frac{7+5+3+4+7+5+7+5+9+6+5}{11} = \frac{63}{11}$$

   $$= 5.7 \text{ (to 1 d.p.)}$$

Skills check:

1. For the data 3, 5, 2, 0, 2, 1, 3, 2, 0, calculate
   a) the mean
   b) the median
   c) the range
   d) the mode.

2. For the data 7, 8, 6, 6, 6, 8, calculate
   a) the mean
   b) the median
   c) the range
   d) the mode.

> Where there is an even number of data values, the median is the average (mean) of the two middle values.

2.1 Averages

Remember
- the mean is the sum of all the values divided by the number of values
- the median is the middle value when the values are arranged in order
- the mode is the most commonly occurring value.

Example 1

A group of students were given a short mental arithmetic test. Their scores were 7, 6, 7, 5, 6, 8, 5, 7, 8, 9.

For this group calculate

a) the mean   b) the median   c) the mode.

a) The sum of the ten scores is 68, so the mean is $\frac{68}{10} = 6.8$.

b) In order, the values are: 5, 5, 6, 6, 7, 7, 7, 8, 8, 9.

    The median is the middle value. Here this is halfway between the 5th and 6th scores, which are both 7, so the median is 7.

c) The score 7 occurs more than any other, so the mode is 7.

Data in a frequency table

The frequency table on the right shows the number of children that 45 women have.

To find the mode, look for the greatest frequency. Here, the category for one child has the greatest frequency, 21. Therefore, the mode is one child.

To find the median, look for the middle value. Here there are 45 values, so the median is the 23rd value, with the values listed in order.

Number of children	Frequency f
0	12
1	21
2	8
3	3
4	0
5	1

Note: The mode is not 21.

Make a running total of the frequencies:
- there are 12 women with no children
- there are 33 women with no children or one child.

The 23rd value, or median, is therefore one child.

To calculate the mean, construct a table of values with a further column showing value × frequency:

Number of children x	Frequency f	xf
0	12	0
1	21	21
2	8	16
3	3	9
4	0	0
5	1	5
	$\sum f = 45$	$\sum xf = 51$

8 women with 2 children each makes 16 children in total.

Write $\sum f$ for 'the sum of the frequencies'.

Write $\sum xf$ for 'the sum of the values'.

The mean number of children is $\dfrac{51}{45}$ = 1.1 (to 1 d.p.).

Data in a grouped frequency table

The ages of the same group of 45 women are shown in the grouped frequency table.

Age (years)	Frequency
15–19	6
20–24	7
25–29	15
30–34	6
35–39	8
40–44	1
45–49	2

The modal class is 25–29 years, as this is the class with the greatest frequency.
The median is the 23rd value, with the values listed in order.

Make a running total of the frequencies:

- there are 6 women less than 20 years old
- there are 6 + 7 = 13 women less than 25 years old
- there are 13 + 15 = 28 women less than 30 years old.

The median age therefore lies in the 25–29 age group.

To calculate the mean you need to add up all the values, but because the frequencies are grouped, you need to use the value at the mid-point of the interval to make an **estimate**:

Age (years)	Frequency f	Mid-interval value m	mf
15–19	6	17.5	105
20–24	7	22.5	157.5
25–29	15	27.5	412.5
30–34	6	32.5	195
35–39	8	37.5	300
40–44	1	42.5	42.5
45–49	2	47.5	95
	$\sum f = 45$		$\sum mf = 1307.5$

Note: Age is a special case; for example, you are 19 until your 20th birthday, so the mid-interval value for the 15–19 age group is 17.5

The estimated mean age of this group of women is therefore $\frac{1307.5}{45} = 29.1$ (to 1 d.p.) – that is, about 29 years old.

You will meet a way of simplifying the arithmetic on pages 27 and 28.

Exercise 2.1

1. A golfer keeps a record of his scores in club competitions during one year. They are

$$75, 77, 77, 74, 79, 76, 84, 76, 75, 77$$

Calculate

a) the modal score

b) the median score

c) the mean score.

2. A vet keeps a record of the number of kittens in litters produced by cats in his practice.

Calculate

a) the modal size of litter

b) the median size of litter

c) the mean size of litter.

Number in litter	Frequency
1	2
2	4
3	7
4	11
5	8
6	4
7	2
8	1

3. A large group of teenagers took part in an exercise session. Their pulse rates were measured before the session began and then measured again after a series of warm-up exercises. The two sets of data are given below.

For each condition, calculate an estimate of the mean pulse rate for the group.

Use your results to compare the distributions.

Pulse rate before warm-up (beats per minute)	Frequency
60–69	5
70–79	8
80–89	22
90–99	29
100–109	13
110–119	5
120–129	0

Pulse rate after warm-up (beats per minute)	Frequency
80–89	9
90–99	12
100–109	25
110–119	17
120–129	9
130–139	7
140–149	3

4. The table shows information about the height of plants of a particular species.

Height (cm)	20–50	50–60	60–65	65–70	70–80	80–90	90–110
Number of plants	27	18	16	15	22	14	14

Calculate an estimate of the mean height of these plants.

5. Students recorded the length of time they spent travelling to school on a particular day.
A summary of the results is shown on the right.

Calculate an estimate of the mean time it took students to travel to school that day.

Time (minutes, correct to the nearest minute)	Frequency
1–15	115
16–25	46
26–35	36
36–55	22
56–80	14

6. The table shows information about the salaries paid to employees in a company.

Salary	Frequency
$0 < x \leq $10\,000$	7
$10\,000 < x \leq $15\,000$	82
$15\,000 < x \leq $20\,000$	45
$20\,000 < x \leq $25\,000$	24
$25\,000 < x \leq $30\,000$	13
$30\,000 < x \leq $50\,000$	4

Calculate an estimate of the mean salary of employees in the company.

2.2 Quartiles and the interquartile range

A useful way of organising data is to divide them into quarters, which are known as **quartiles**.

These dots represent 19 values:

•• ••• •• •••• •• ••• •••

Q_1 is the **lower quartile**.

For a data set with n values $(x_1, x_2, ..., x_n)$, calculate $\frac{1}{4}n$.

If $\frac{1}{4}n$ is an integer r then Q_1 is the mid-point of x_r and x_{r+1}.

If $\frac{1}{4}n$ lies between r and $r + 1$ then Q_1 is x_{r+1}.

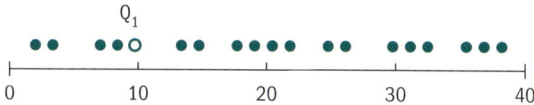

$$Q_1$$
•• ••○ •• •••• •• ••• •••
├─────┼─────┼─────┼─────┤
0   10   20   30   40

$\frac{1}{4} \times 19 = 4.75$, so Q_1 is the 5th value.

Q_2 is the **median**.

To find the median, calculate $\frac{1}{2}n$.

If $\frac{1}{2}n$ is an integer r, then Q_2 is the mid-point of x_r and x_{r+1}.

If $\frac{1}{2}n$ lies between r and $r + 1$, then Q_2 is x_{r+1}.

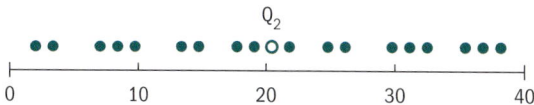

$$Q_2$$
•• ••• •• ••○• •• ••• •••
├─────┼─────┼─────┼─────┤
0   10   20   30   40

$\frac{1}{2} \times 19 = 9.5$, so Q_2 is the 10th value.

Q_3 is the **upper quartile**.

To find the upper quartile, calculate $\frac{3}{4}n$.

If $\frac{3}{4}n$ is an integer r, then Q_3 is the mid-point of x_r and x_{r+1}.

If $\frac{3}{4}n$ lies between r and $r+1$, then Q_3 is x_{r+1}.

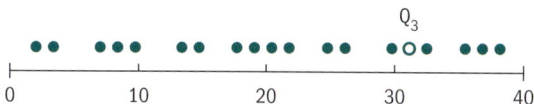

$$Q_3$$
•• ••• •• •••• •• •○• •••
├─────┼─────┼─────┼─────┤
0   10   20   30   40

$\frac{3}{4} \times 19 = 14.25$, so Q_3 is the 15th value.

> The **range** $= Q_4 - Q_0$
> The range indicates the spread of the values.

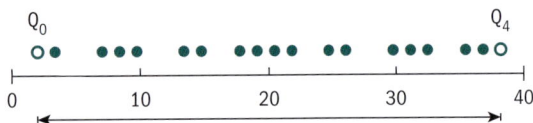

> The **interquartile range** (IQR) $= Q_3 - Q_1$
> The IQR indicates the spread of the middle 50 per cent of values.

Note: The IQR is the measure of spread always associated with the median, which is a measure of centre.

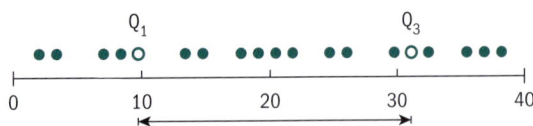

The semi-interquartile range is simply half of the IQR, and is sometimes used as an alternative measure of spread.

Example 2

The scores of a group of students in a short mental arithmetic test (from Example 1) were, in order,

5, 5, 6, 6, 7, 7, 7, 8, 8, 9

Calculate

a) the quartiles **b)** the interquartile range.

. . . .

a) $n = 10$, so $\frac{1}{4}n = 2.5$ and Q_1 is the 3rd value in the ordered list. $Q_1 = 6$

$\frac{1}{2}n = 5$, so Q_2 is the mid-point of the 5th and 6th values in the ordered list.

These are both 7, so the median, $Q_2, = 7$.

$\frac{3}{4}n = 7.5$ so Q_3 is the 8th value in the ordered list. $Q_3 = 8$.

b) IQR $= 8 - 6 = 2$

Exercise 2.2

1. A golfer keeps a record of his scores in club competitions during one year. They are

75, 77, 77, 74, 79, 76, 84, 76, 75, 77

Calculate the lower quartile and the upper quartile for these data.

In Exercise 2.1, you calculated the medians of the data sets in questions 1 and 2.

2. A vet keeps a record of the number of kittens in litters produced by cats in his practice.

Calculate the lower quartile and the upper quartile for these data shown in the table on the right.

Number in litter	Frequency
1	2
2	4
3	7
4	11
5	8
6	4
7	2
8	1

3. A group of students recorded their pulse rate when they were in a relaxed, restful state.

Their results (in beats per minute), were:

53 54 55 55 57 58 59 59 59 62 62 62

62 62 63 63 63 64 64 65 65 67 67 67

68 68 69 69 70 71 71 73 75 79 80 83

Calculate

a) the median

b) the interquartile range of these pulse rates.

4. The ages of a group of 45 women are:

17 17 17 19 19 19 20 22 22 22 23 23 24 25 25

25 26 26 26 26 27 27 28 28 28 29 29 29 30 32

32 32 34 34 35 35 35 35 36 37 38 38 44 45 48

Calculate

a) the median

b) the interquartile range of their ages.

5. At a metro station, a regular passenger records how long (in seconds) he has to wait for a train to arrive once he gets to the platform. His results are:

87   42   0   62   124   0   58   37   74   94

182   23   17   62   29   17   82   54   0   45

Find

a) the median

b) the interquartile range of these times.

6. The length of time, in minutes, that customers spend in a coffee shop is recorded. The results are:

17, 15, 9, 31, 33, 41, 8, 14, 13, 22, 27, 43, 32, 14

Find

a) the median

b) the interquartile range of these times.

2.3 Variance and standard deviation

The mean of a set of data is given by the formula

$$\text{Mean} = \bar{x} = \frac{\sum x}{n} \quad \text{or} \quad \frac{\sum xf}{\sum f}$$

Where x is each data value, n is the number of data values and f is the frequency.

Remember the greek letter sigma, $\sum$, stands for 'the sum of'.

So far you have met two measures of spread: the range and the interquartile range. Each of these measures uses only two values.

Another measure of spread, one which uses all the data values, is the **variance**.

The variance of a set of data is defined as the average of the squared distances from the mean.

$$\text{Variance} = \sum \frac{(x - \bar{x})^2}{n} \quad \text{or} \quad \frac{\sum (x - \bar{x})^2 f}{\sum f}$$

$$= \frac{\sum x^2}{n} - \bar{x}^2 \quad \text{or} \quad \frac{\sum x^2 f}{\sum f} - \bar{x}^2$$

This second version of the formula is generally easiest to work with.

The variance is normally denoted by σ^2.

To show that the first equation is equal to the second, one can be derived from the other:

$$\sum \frac{(x - \bar{x})^2}{n} = \sum \left(\frac{x^2 - 2x\bar{x} + \bar{x}^2}{n} \right)$$

$$= \sum \left(\frac{x^2}{n} \right) - 2\bar{x} \sum \left(\frac{x}{n} \right) + \bar{x}^2$$

$$= \sum \left(\frac{x^2}{n} \right) - 2\bar{x}^2 + \bar{x}^2$$

$$= \frac{\sum x^2}{n} - \bar{x}^2$$

The **standard deviation** is the square root of the variance.

It is a measure of how spread out the set of data is, and it is in the same units as the data.

σ is used to denote standard deviation.

Example 3

Find the mean and standard deviation of this set of data

11, 13, 14, 16, 18

$\sum x = 11 + 13 + 14 + 16 + 18 = 72$

so the mean, $\bar{x} = \dfrac{72}{5} = 14.4$

$\sum x^2 = 11^2 + 13^2 + 14^2 + 16^2 + 18^2 = 1066$

so the variance, $\sigma^2 = \dfrac{1066}{5} - 14.4^2 = 5.84$

and the standard deviation, σ
$= \sqrt{5.84} = 2.42$ (to 3 s.f.).

It is also possible to use the second form of the variance formula:

$\sum (x - \bar{x})^2 = (11 - 14.4)^2 + (13 - 14.4)^2 + (14 - 14.4)^2 + (16 - 14.4)^2 + (18 - 14.4)^2$

$\sum (x - \bar{x})^2 = (^-3.4)^2 + (^-1.4)^2 + (^-0.4)^2 + 1.6^2 + 3.6^2 = 29.2$

so the variance, $\sigma^2, = \dfrac{29.2}{5} = 5.84$

> You can use your calculator's statistical functions to get the mean and standard deviation. Your calculator will normally have two versions, which give slightly different answers.

The **population standard deviation** is the one that you want, and is usually shown as σ or σ_n on a calculator.

The other version is sometimes referred to as a **sample standard deviation**, and is usually shown as s or σ_{n-1}, but you don't need to work with it at the moment.

Only one of these is dealt with so that σ can be used without confusion.

Example 4

The number of children, x, that each of a class of 30 students has in their family was recorded.

Find the mean and standard deviation of the number of children in a family for this set of data, given that $\sum x = 66$ and $\sum x^2 = 165$.

Mean, $\bar{x}, = \dfrac{66}{30} = 2.2$

Variance, $\sigma^2, = \dfrac{165}{30} - 2.2^2 = 0.66$

So standard deviation, $\sigma, = \sqrt{0.66} = 0.812$ (to 3 s.f.).

Example 5

The number of putts, x, a golfer takes on each of the 18 holes in a round is recorded. Find the standard deviation, σ, of the number of putts he takes on each hole, given that $\sum (x - \bar{x})^2 = 4.944$.

There were 18 holes, so variance, $\sigma^2, = \dfrac{4.944}{18} = 0.2746\ldots$

and standard deviation, $\sigma, = \sqrt{0.2746\ldots} = 0.524$ (to 3 s.f.).

Example 6

The heights of 142 plants are recorded in the table, where, for example, the class 65–70 means at least 65 cm and less than 70 cm tall.

Class	55–60	60–65	65–70	70–75	75–80	80–85	85–90
Frequency	2	11	37	54	28	9	1

Calculate estimates of the mean, $\bar{x}$, and variance, σ^2, for the height of the plants.

You need to calculate the mid-point of each class, and a table of values is the simplest way to do this:

Class	Mid-point m	Frequency f	mf	m^2f
55–60	57.5	2	115	6612.5
60–65	62.5	11	687.5	42 968.75
65–70	67.5	37	2497.5	168 581.3
70–75	72.5	54	3915	283 837.5
75–80	77.5	28	2170	168 175
80–85	82.5	9	742.5	61 256.25
85–90	87.5	1	87.5	7656.25
		$\sum f = 142$	$\sum mf = 10\,215$	$\sum m^2 f = 739\,087.5$

So $\bar{x} = \dfrac{10\,215}{142} = 71.9$ (to 3 s.f.)

and variance, $\sigma^2, = \dfrac{739\,087.5}{142} - \left(\dfrac{10\,215}{142}\right)^2 = 30.0$ (to 3 s.f.).

What does 'standard deviation' actually mean?

A good visual idea of standard deviation is to imagine a histogram of the data. For a roughly symmetrical distribution, a spread of 4 standard deviations covers approximately the central 95% of the distribution.

> You will learn about a very important symmetrical distribution in Chapter 8.

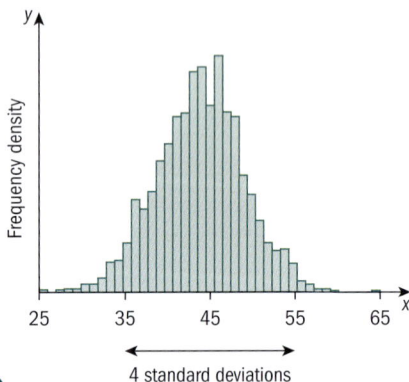

4 standard deviations

Exercise 2.3

1. For each set of data, calculate the mean and standard deviation. You should do this using the formulae, and check using your calculator.

   a) 12   17   11   8   6   18   14   17

      11   15   16   18   9   15   20   14

   b)

x	4	5	6	7	8	9
f	12	18	35	28	16	9

2. The wingspans of 352 Great Tits were measured. The information is summarized in the table.

Wingspan (cm)	70	71	72	73	74	75	76	77	78	79
Frequency	4	19	53	77	23	75	53	30	12	6

   Calculate the mean and standard deviation of the wingspan of a Great Tit.

3. The African grey parrot are an endangered species of small birds found in Africa. The weights (in grams, to the nearest gram) of 210 African grey parrots were measured during 2014. The information is summarized in the table below.

Weight (grams)	300–319	320–339	340–359	360–379	380–399
Frequency	3	45	149	11	2

   Calculate estimates of the mean and standard deviation of the weight of an African grey parrot.

4. $\sum x = 75$, $\sum x^2 = 293$, $n = 30$

   Find the mean and variance of X.

5. $\sum x = 2.1$, $\sum x^2 = 83.1$, $n = 7$

   Find the mean and standard deviation of X.

6. $\sum (x - \bar{x})^2 = 44.3$, $n = 16$

   Find the variance of X.

7. $\sum (x - \bar{x})^2 = 19\ 735.4$, $n = 37$

Find the standard deviation of X.

8. Estimate standard deviation of this distribution

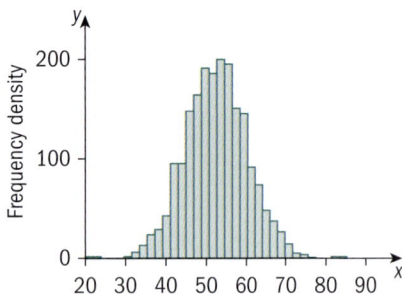

2.4 Which average should you use?

The simple answer is that if one measure was always the best then there would be no need for any others – so it depends!

The mode is referred to as an average, but it isn't the same sort of measure as the median and mean, and if you could only have one of these three, the mode is not the one you would choose.

The median is the average – half of the other values are bigger and half are smaller – while the mean is the value which gives everyone an equal share. This suggests a strategy for deciding which measure to use: if the total of the values tells you something of interest, then the mean is likely to be the more appropriate. Note that it needs to be the *total* that tells you something of interest: if you were looking at the wages paid in a factory because you were thinking of applying for a job there, the median would be the right average to look at, since it would tell you what a typical worker is earning; on the other hand, if you wanted to buy the factory then the total wage bill would be of interest, since it tells you about running costs, and therefore the mean would be the more appropriate.

So the most appropriate average is not simply a matter of whether the data are symmetrical, or whether there are outliers; using a set of data for different purposes means different averages are appropriate. If the data are roughly symmetrical then the median and mean will be close together anyway, so it won't matter which you calculate – which tells you that symmetry is not a good criterion for choosing which to use.

There is, however, one golden rule: the measures of centre and spread are *not* interchangeable! The median goes with the interquartile range and the mean goes with the standard deviation (or variance).

2.5 Coding

The diagram shows two sets of data:
Dataset 1 represents the ages, in years, of a group of people; Dataset 2 represents the ages of the same people 15 years later.

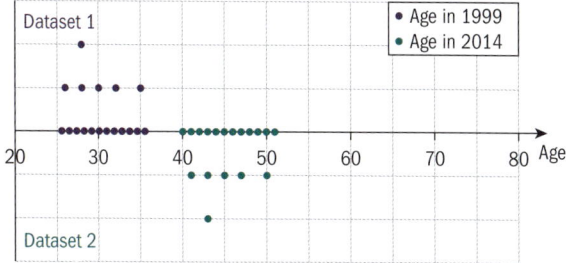

All the data values have shifted by 15, so

Mean of Dataset 2 = (mean of Dataset 1) + 15

and there is no change to any measure of spread.

The next diagram shows a set of temperatures in degrees Celsius (Dataset 1) and the same temperatures in degrees Fahrenheit (Dataset 2).

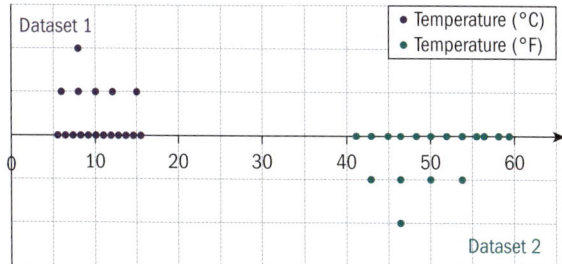

To get the values in degrees Fahrenheit, all of the values in degrees Celsius have been multiplied by 1.8, then 32 added:

Mean of Dataset 2 = 1.8 × (mean of Dataset 1) + 32

The degree of spread is unaffected by the + 32.

Standard deviation of Dataset 2 = 1.8 × (standard deviation of Dataset 1)
So variance of Dataset 2 = 1.8^2 × (variance of Dataset 1)

If a set of data values X is related to a set of values Y so that $Y = aX + b$, then:

mean of $Y = a \times$ mean of $X + b$

standard deviation of $Y = a \times$ standard deviation of X

variance of $Y = a^2 \times$ variance of X.

You can use this idea to transform or **code** a set of numbers. You might do this to make the numbers easier to work with.

Another use of coding is in standardizing different data sets for comparison.

Example 7

Here is the grouped data from Example 6, on the heights of 142 plants. (Recall that the class 65–70, for example, means at least 65 cm and less than 70 cm tall.)

Use the coding $M = \dfrac{m-57.5}{5}$ to find the mean and the variance of these data.

Class	m	f
55–60	57.5	2
60–65	62.5	11
65–70	67.5	37
70–75	72.5	54
75–80	77.5	28
80–85	82.5	9
85–90	87.5	1
		$\sum f = 142$

Extend the table:

Class	m	M	f	Mf	M²f
55–60	57.5	0	2	0	0
60–65	62.5	1	11	11	11
65–70	67.5	2	37	74	148
70–75	72.5	3	54	162	486
75–80	77.5	4	28	112	448
80–85	82.5	5	9	45	225
85–90	87.5	6	1	6	36
			$\sum f = 142$	$\sum Mf = 410$	$\sum M^2 f = 1354$

The coded numbers M are easier than the uncoded numbers m

Then $\bar{M} = \dfrac{410}{142} = 2.887\ldots$ and $\sigma_M^2 = \dfrac{1354}{142} - \bar{M}^2 = 1.19857\ldots = 1.20$ (to 3 s.f.)

To find the mean and variance of the uncoded data:

$\bar{m} = \bar{M} \times 5 + 57.5 = 2.887 \times 5 + 57.5 = 71.9$ (to 3 s.f.)

$\sigma_m^2 = 5^2 \times \sigma_M^2 = 25 \times 1.19857\ldots = 30.0$ (to 3 s.f.)

Without coding:

$\sum mf = 10\,215$

$\sum m^2 f = 739\,087.5$

$\bar{m} = \dfrac{10\,215}{142} = 71.9$ (to 3 s.f.)

$\sigma_m^2 = \dfrac{739\,087.5}{142} - \left(\dfrac{10\,215}{142}\right)^2 = 30.0$ (to 3 s.f.)

Coding gives the same answers, but uses smaller numbers and fewer key presses

Exercise 2.4

1. A set of observations $\{x\}$ are coded using $X = \frac{x - 62.5}{5}$. $\bar{X} = 3.6$, $\text{Var}(X) = 5.3$.

   Calculate the mean and variance of the original set of observations.

2. A set of observations $\{x\}$ are coded using $X = \frac{x - 1055}{10}$. $\bar{X} = 8.2$, $\text{Var}(X) = 3.22$.

   Calculate the mean and variance of the original set of observations.

3. The temperatures (in °F) at a resort are measured at the same time on eight successive Mondays:

   50.0   53.6   52.7   55.4   55.4   57.2   59.9   62.6

   a) Calculate the mean of these temperatures (in °F).

   b) Convert each temperature into °C using the formula $C = \frac{F - 32}{\frac{9}{5}}$.

   c) Calculate the mean of these temperatures (in °C).

   d) Check that the mean temperature in °F converts to the same mean in °C.

4. The table shows information about the time taken by a bus to travel from Nikki's home to her grandmother's house.

Time (minutes)	Frequency
$8 \leq x < 10$	5
$10 \leq x < 12$	14
$12 \leq x < 14$	3
$14 \leq x < 16$	2

   a) Calculate estimates of the mean and variance of the time taken by the bus.

   b) Code this data set using $X = \frac{x - 9}{2}$.

   c) Find estimates of the mean and variance of the coded data.

   d) Check that the mean of x is 9 more than twice the mean of X, and the variance of x is four times the variance of X.

5. The table shows information about the salaries paid to a company's employees.

Salary	Frequency
$0 < x \le \$10\,000$	12
$\$10\,000 < x \le \$15\,000$	63
$\$15\,000 < x \le \$20\,000$	32
$\$20\,000 < x \le \$25\,000$	21
$\$25\,000 < x \le \$30\,000$	8
$\$30\,000 < x \le \$50\,000$	4

a) Code these data using $X = \dfrac{x - 5000}{2500}$.

b) Calculate estimates of the mean and variance of the coded data.

c) Use your answers to part **(b)** to give estimates of the mean and variance of the salaries in the company.

6. The table shows information about the daily time spent by a nurse with each patient.

Time, t (minutes)	Frequency
$10 < t \le 15$	312
$15 < t \le 20$	479
$20 < t \le 25$	243
$25 < t \le 30$	119

a) Code these data using $T = \dfrac{t - 12.5}{5}$.

b) Calculate estimates of the mean and variance of the coded data.

c) Use your answers to **(b)** to give estimates of the mean and variance of the daily time spent by the nurse with each patient.

Summary exercise 2

1. $\sum x = 150$ $\sum x^2 = 1535$ $n = 25$

   Find the mean and variance of X.

2. A summary of 20 observations of y gave the following information

   $\sum(y-a) = -37$ $\sum(y-a)^2 = 1529$

   Find the mean and standard deviation of y.

3. The weights (in kg) of the hand luggage carried on to a flight by 97 passengers are summarised by

   $\sum(x-5) = 314$ $\sum(x-5)^2 = 1623$.

   Find $\sum x$ and $\sum x^2$.

4. The test score, x, for a class of 14 students gives $\sum x = 919$ and $\sum x^2 = 60\,773$.

   a) Calculate the mean and variance of the marks for this class.

      Another class of 15 students taking the same test scored a mean of 63.8, with a standard deviation of 5.58 marks.

   b) Calculate $\sum x^2$ for the second class.

   c) Calculate the mean mark for all of the students in the two classes.

5. An airport bus service runs from the city centre and from the airport every 10 minutes during the day. The number of passengers on a random sample of journeys from the airport is shown below.

   11, 3, 14, 34, 1, 6, 12, 15, 8, 4, 28, 7, 3, 0, 8, 12

   Calculate the median and interquartile range of the number of passengers.

6. Students' marks on a history examination were: 73, 67, 76, 63, 59.

   a) Calculate the mean and standard deviation of these marks.

      For comparisons with other subjects, the marks are to be scaled so that they have a mean of 50 and a standard deviation of 10. This is to be done by using the transformation $Y = \dfrac{X-a}{b}$, where X is the original mark and Y is the transformed mark.

   b) Calculate the values of a and b.

7. A summary of 24 observations of x gave the following information:

   $\sum(x-a) = -73.2$ and $\sum(x-a)^2 = 2115$.

   The mean of these values of x is 8.95.

   i) Find the value of the constant a.   [2]

   ii) Find the standard deviation of these values of x.   [2]

   Cambridge International AS and A Level Mathematics 9709, Q1 Paper 6 October/November 2007

8. The salaries, in thousands of dollars, of 11 people, chosen at random in a certain office, were found to be:

   40, 42, 45, 41, 352, 40, 50, 48, 51, 49, 47.

   Choose and calculate an appropriate measure of central tendency (mean, mode or median) to summarise these salaries. Explain briefly why the other measures are not suitable.   [3]

   Cambridge International AS and A Level Mathematics 9709 Q1 Paper 6 May/June 2006

9. The following table shows the results of a survey to find the average daily time, in minutes, that a group of schoolchildren spent in internet chat rooms.

Times per day (t minutes)	Frequency
$0 \leq t < 10$	2
$10 \leq t < 20$	f
$20 \leq t < 40$	11
$40 \leq t < 80$	4

The mean time was calculated to be 27.5 minutes.

i) Form an equation involving f and hence show that the total number of children in the survey was 26. [4]

ii) Find the standard deviation of these times. [2]

Cambridge International AS and A Level Mathematics 9709 Q2 Paper 6 May/June 2005

10. Rachel measured the lengths in millimetres of some of the leaves on a tree. Her results are recorded below.

32 35 45 37 38 44 33 39 36 45

Find the mean and standard deviation of the lengths of these leaves. [3]

Cambridge International AS and A Level Mathematics 9709 Q1 Paper 6 November 2008

11. The amounts of money, x dollars, that 24 people had in their pockets are summarised by $\sum(x - 36) = -60$ and $\sum(x - 36)^2 = 227.76$. Find $\sum x$ and $\sum x^2$. [5]

Cambridge International AS and A Level Mathematics 9709 Q2 Paper 61 October/November 2012

12. A summary of 30 values of x gave the following information:

$$\sum(x - c) = 234, \quad \sum(x - c)^2 = 1957.5,$$

where c is a constant.

i) Find the standard deviation of these values of x. [2]

ii) Given that the mean of these values is 86, find the value of c. [2]

Cambridge International AS and A Level Mathematics 9709, Paper 61 Q1 May/June 2013

13. Swati measured the lengths, x cm, of 18 stick insects and found that $\sum x^2 = 967$. Given that the mean length is $\frac{58}{9}$ cm, find the values of $\sum(x - 5)$ and $\sum(x - 5)^2$. [5]

Cambridge International AS and A Level Mathematics 9709, Paper 61 Q3 October/November 2013

14. The length of time, t minutes, taken to do the crossword in a certain newspaper was observed on 12 occasions. The results are summarised below.

$$\sum(t - 35) = -15 \quad \sum(t - 35)^2 = 82.23$$

Calculate the mean and standard deviation of these times taken to do the crossword. [4]

Cambridge International AS and A Level Mathematics 9709, Paper 6 Q1 May/June 2007

Chapter summary

- Q_2 is the **median**.

  To find the median, calculate $\frac{1}{2}n$.

  If $\frac{1}{2}n$ is an integer r then Q_2 is the mid-point of x_r and x_{r+1}.

  If $\frac{1}{2}n$ lies between r and $r+1$ then Q_2 is x_{r+1}.

- Q_1 is the **lower quartile** and Q_3 is the **upper quartile**.

  For a data set with n values $(x_1, x_2, \ldots, x_n)$, to find Q_1, calculate $\frac{1}{4}n$.

  If $\frac{1}{4}n$ is an integer r, then Q_1 is the mid-point of x_r and x_{r+1}.

  If $\frac{1}{4}n$ lies between r and $r+1$, then Q_1 is x_{r+1}.

- The same process is used to find Q_3, but using $\frac{3}{4}n$ instead.

- The **interquartile range** (IQR) is the difference between the lower and upper quartiles:

  $IQR = Q_3 - Q_1$.

- For a set of data,

  $$\text{Mean} \quad = \bar{x} = \frac{\sum x}{n} \quad \text{or} \quad \frac{\sum xf}{\sum f}$$

  and

  $$\text{Variance} \quad = \sum \frac{(x - \bar{x})^2}{n} \quad \text{or} \quad \frac{\sum (x - \bar{x})^2 f}{\sum f}$$

  $$= \frac{\sum x^2}{n} - \bar{x}^2 \quad \text{or} \quad \frac{\sum x^2 f}{\sum f} - \bar{x}^2$$

  For grouped data, the mid-point of the interval is used for x in each case.

- The **standard deviation** is the square root of the variance and is a measure of how spread out the data are. It is measured in the same units as the data.

If a set of data values X is related to a set of values Y so that $Y = aX + b$, then: mean of $Y = a \times$ mean of $X + b$, standard deviation of $Y = a \times$ standard deviation of X, and variance of $Y = a^2 \times$ variance of X.

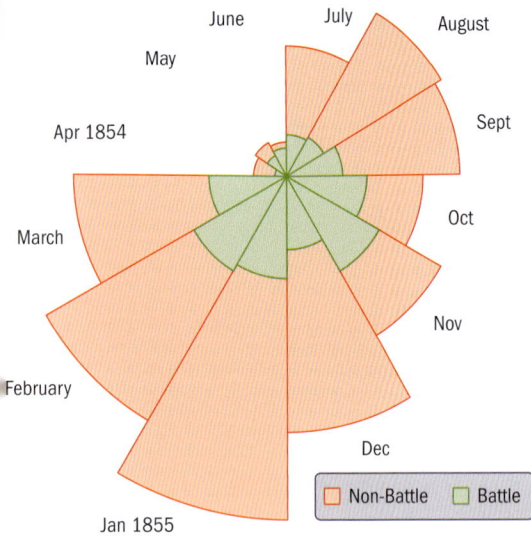

Florence Nightingale worked as a nurse in the Crimean war despite being born into a wealthy British family. She is famous for the reforming hygiene in hospitals when she discovered that more soldiers were dying from diseases in the hospitals than from their battle wounds. It was her success in communicating these findings by using innovative statistical diagrams (coxcombs – the first examples of pie charts) which gave her discoveries the political capital to make a series of fundamental health and hygiene reforms. As a result, she was the first woman to be elected to the Royal Statistical Society, and she was the first woman to receive the Order of Merit which is Britain's highest civilian honour.

Objectives

- Select a suitable way of presenting raw statistical data, and discuss advantages and or disadvantages that particular representations may have.
- Construct and interpret stem-and-leaf diagrams, box-and-whisker plots, histograms and cumulative frequency graphs.
- Use a cumulative frequency graph to estimate the median value, the quartiles and the inter-quartile range of a set of data.

Before you start

You should know how to:

1. Find the median and quartiles of a data set given as a list, a frequency table and a grouped frequency table, e.g. For the data sets given find the median and quartiles.

   a) 7, 7, 8, 9, 10, 12, 14

   7, 7, 8, **9**, 10, **12**, 14 – the median is the middle value (9), the lower quartile (7) is the middle value of the bottom half, and the upper quartile (12) is the middle of the top half.

Skills check:

1. For the data sets given find the median and quartiles.

   a) 8, 8, 9, 10, 11, 12, 16, 18, 19, 22, 23

   b)

Number of correct answers in short test	3	4	5	6	7
Frequency	4	7	6	3	1

b)

Number of beans per pod	1	2	3	4	5
Frequency	2	5	8	6	4

There are 2 + 5 + 8 + 6 + 4 = 25 values summarized in the table so the median is the 13th, which is a 3, the Lower quartile is the 7th, which is a 2, the Upper quartile is the 19th, which is a 4.

c)

Weight of sample of birds (to nearest gram)	Frequency
9–12	15
13–20	20
21–29	28
30–40	16
41–55	25

3.1 Stem-and-leaf diagrams

Stem-and-leaf diagrams give the shape of a distribution in the same way as a histogram with equal intervals does, but also keep the detail available.

> This means you will be able to draw a box-and-whisker plot – see section 3.2.

The ordering in the stem-and-leaf diagram makes it easy to find the median and quartiles, as well as the highest and lowest values.

The stem-and-leaf diagram below shows the heights of some plants:

```
3 | 5  7  7              (3)
4 | 2  2  3  4  7        (5)
5 | 1  4  7  8           (4)
6 | 2                    (1)
        key   5 | 2   means 52 cm
```

There are 13 plants, so the median is the 7th – shown in red (44 cm).

Construct a stem-and-leaf in two stages:
(1) First we assign the values to the correct place on the stem
(2) Then order each 'leaf' in turn.

The data in this diagram might have been provided exactly as it was collected: as a simple unordered list of data.

$$42, 57, 37, 62, 51, 42, 47, 35, 54, 43, 44, 58, 37$$

In this case, the first stage the diagram will be:

```
3 | 7  5  7
4 | 2  2  7  3  4
5 | 7  1  4  8
6 | 2
```

> There are two other things to note with the final output. The key is compulsory – and it needs to include any units that are used in the measurement. The numbers in brackets at the end of each row are optional, but make finding the median and quartiles easier.

Then it is easy to order each row of data to get the final stem-and-leaf diagram.

3	5	7	7			(3)
4	2	2	3	4	7	(5)
5	1	4	7	8		(4)
6	2					(1)

key   5 | 2   means 52 cm

In Chapter 2, you learned how to find the quartiles and inter-quartile range for a set of data – here the lower quartile (42 cm) and upper quartile (54 cm) are highlighted in blue, giving an inter-quartile range of 12 cm.

Two sets of related data can be shown as a back-to-back stem-and-leaf diagram, where a common stem is used and the two sets of data go to the left and right.

> Back-to-back stem-and-leaf diagrams allow comparison of the distributions to be made.

You should make a comment about the location and spread of the two, but try to describe them in context rather than just list values of medians and inter-quartile range, e.g. if another type of plant had heights which were all in the 50–70 cm range you could say that this type tended to be smaller and were more variable in height than the second type of plant.

Example 1

The heights of samples of two types of plants are represented in this stem-and-leaf diagram. Compare the heights of the two types of plants.

		Type A								Type B					
						2	3	5	7	7				(4)
(2)					7	4	3	2	2	4	8	8	9	(6)
(6)	9	7	7	6	3	1	4	1	4	5	5	7	(5)
(5)		8	5	4	4	2	5	2	3				(2)
(3)			5	3	3	6							

Key   1 | 4 | 5   means 41 cm for type A and 45 cm for type B

The median of type A is 50.5 cm and of type B it is 38. The IQR for A is 12 and for B it is 13.

Plants of type A are taller on average than plants of type B, and the two types of plant have similar variability in their heights.

> When comparing distributions, you should comment on average and spread. The comments need to be in the context of the data.

The number of minor mistakes on driving tests marked by an examiner (A) are listed below.

12	8	11	14	18	14	15	11
9	7	12	18	11	17	16	10

Stem-and-leaf diagrams normally have only one set of leaves for any stem value but either two or five sets of leaves are possible (five sets is very uncommon).

Using two stems with leaves 0–4 and 5–9 and assigning each data value to a stem.

```
0 | 8  9  7
1 | 2  1  4  4  1  2  1  0
1 | 8  5  8  7  6
```

Then ordering the leaves gives the final stem-and-leaf diagram.

```
0 | 7  8  9
1 | 0  1  1  1  2  2  4  4
1 | 5  6  7  8  8
```

Key: 1 | 6   means 16 minor mistakes.

The stem-and-leaf diagram shows the number of minor mistakes made on driving tests marked by another examiner (B).

```
0 | 1  4                          (2)
0 | 5  7  7                       (3)
1 | 0  0  1  2  3  3  4  4        (8)
1 | 6  7  8  8                    (4)
2 | 0  1                          (2)
```

Key: 1 | 6   means 16 minor mistakes.

```
            Examiner A                              Examiner B
                                    0 | 1  4                                   (2)
(3)                      9  8  7    0 | 5  7  7                                (3)
(8)    4  4  2  2  1  1  1  0       1 | 0  0  1  2  3  3  4  4                 (8)
(5)                8  8  7  6  5    1 | 6  7  8  8                            (4)
                                    2 | 0  1                                   (2)
```

Key: 2 | 1 | 5   means 15 minor mistakes for Examiner B
                     and 12 minor mistakes for Examiner A

The two examiners seem to identify approximately the same number of minor mistakes on average, however examiner B has a much wider range.

Note that unless you know something more about the candidates it is not possible to know whether this is a difference in marking styles or because the groups were not of comparable standard.

Exercise 3.1

1. The data below records the masses, in grams, of samples of 35 plums of two different types.

	Type A				Type B	
(1)	2	5		(1)	4	4
(3)	3	0 2 4		(11)	4	5 6 7 7 8 8 8 9 9 9 9
(5)	3	5 5 7 7 9		(14)	5	0 0 0 0 1 1 1 1 1 2 2 3 3 4
(9)	4	1 2 2 2 2 4 4 4 4		(9)	5	5 5 6 6 7 7 8 9 9
(10)	4	5 5 6 6 6 7 7 7 7 8				
(4)	5	0 1 2 2				
(3)	5	7 8 9				

Key: 4 | 5   means 45 grams for each diagram

a) For each type, find the median and inter-quartile range of the masses.
b) Draw a back-to-back stem-and-leaf diagram to show these data in a single diagram.
c) Compare the characteristics of the two types of plums.
d) If you were a fruit-grower, which of the two types would you plant? Give a reason.

2. A group of students measured their pulse rates at the start of a statistics lesson. They then did 5 minutes of moderate exercise and measured their pulse rates again. The results are summarised below:

			Before exercise			After exercise						
(3)			9	8	6	6						
(4)		7	5	4	2	7						
(4)		8	6	5	1	8	4	5	7	7	8	(5)
(3)			5	3	2	9	1	3	4	4	6 9	(6)
(1)					0	10	0	2	5	5		(4)

Key: 1 | 9 | 5   means 91 beats per minute before exercise
and 95 beats per minute after exercise

Compare the pulse rates of the group before and after exercise.

3. Samples of the weights of Dunnocks, sometimes known as Hedge Sparrows, are taken in January (during the winter) and April (when their breeding season starts).

	January		April	
		18	6	(1)
(1)	5	19	1 4 5 7 8	(5)
(4)	1 2 6 9	20	0 2 4 4 5 8 9	(7)
(7)	0 2 5 5 8 8 9	21	0 1 2 5 5 5 6 7 8 9	(10)
(8)	2 2 3 3 3 5 6 7	22	0 0 2 3 3 4 8 9	(8)
(8)	0 1 2 4 4 9 9 9	23	4	(1)
(5)	0 1 2 3 4	24	5 7	(2)
(1)	2	25		
(2)	0 2	26		

Key: 7 | 22 | 5   means 22.7 grams in January and 22.5 grams in April

Compare the weights of Dunnocks in January and in April.

4. The wingspans of a sample of male and female Eurasian Eagle-Owls were collected, and are summarized in the diagram below

		Male					Female				
(2)				9	7	15					
(5)	8	6	6	3	2	16	8	9			(2)
(6)	7	6	6	5	3	1	17	0 3 5 5 7			(5)
(1)					3	18	2 5 7 9 9				(5)
						19	0 3				(2)

Key: 1 | 17 | 5   means 1711 cm for Males and 175 cm for Females

Compare the wingspans of male and female Eurasian Eagle-Owls.

5. The number of grams of carbohydrates in a typical portion of some different fruits and vegetables are given. Construct a back-to-back stem-and-leaf diagram and compare the amount of carbohydrates in fruit and vegetables.

Fruit

34   3   30   12   23   12   20   19
15   26   13   19   11   26   21

Vegetables

4   6   8   7   5   4   2   5
2   2   3   11   26   18   5

3.2 Box-and-whisker plots

Box-and-whisker plots typically show only five 'points' from the distribution - the top and bottom of the range, the median and the upper and lower quartiles.

Note: There is no detail to distract your attention from the big picture.

These 5 values are sometimes known as the quartiles and the notation Q_0, Q_1, Q_2, Q_3 and Q_4 is sometimes used to refer to minimum, LQ, median, UQ and maximum respectively.

The pair of box-and-whisker on the right show the heights of two groups of plants. We see that group 1 are much more consistent in their heights than group 2, and group 2 tend to be taller.

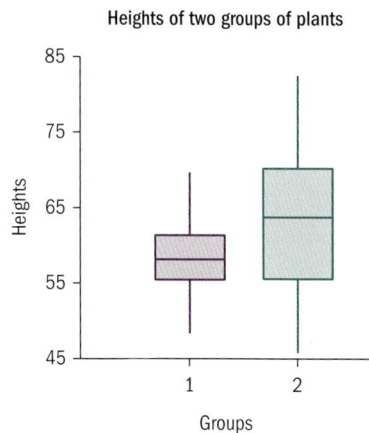

Heights of two groups of plants

The decathlon is a set of 10 running, jumping and throwing events in which athletes are awarded points in each event based on their performance. The total points across the 10 events determines the result of the decathlon competition. The three box-and-whisker below show the scores across the 10 events for the two best decathlons ever (up to June 2014) which scored 9039 and 9026 points and the personal best of the 50th best decathlete (8506 points).

The strengths and weaknesses of the box-and-whisker are vividly illustrated by this example. The 'average' in the top two is very similar – the middle of the box – but Roman Sebrle's performances across the 10 events were much more consistent that Ashton Eaton's, while Valter Külvet's performances are even more consistent but at a much lower level than the other two. The weakness is that you have no detail beyond the five summary statistics – and in this case that means you cannot make any comparison of performances in individual events.

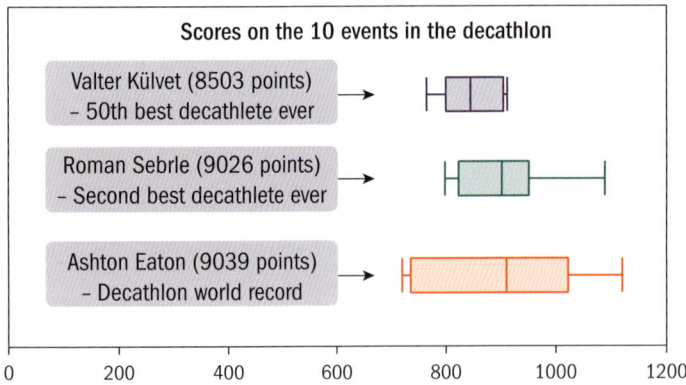

Scores on the 10 events in the decathlon

Outliers*

Outliers are values that are uncommonly large or small for the data. One common definition of an outlier is a value which is more than $1.5 \times IQR$ above the UQ or below the LQ – these limits are known as *fences*.

$$x - Q_3 > 1.5 \times (Q_3 - Q_1)$$
$$Q_1 - x > 1.5 \times (Q_3 - Q_1)$$

Note: This section is not required for the Cambridge 9709 syllabus, but it may aid in your understanding.

We can see that there are no outliers for any of the three decathlons shown above – none of the whiskers extend far enough to be 1.5 times the width of the box. Question 2 below represents outliers by *s beyond the main whisker – allowing you see to see the bulk of the distribution and any unusually large or small values individually. Because of the amount of work involved in calculating what is an outlier and in plotting, outliers are not part of this course, but computers can be programmed to plot them automatically, and you should be familiar with the ideas.

Exercise 3.2

1. The box-and-whisker plots below show the average daily temperatures in two holiday resorts in July. Compare the two resorts as holiday destinations, and comment on other information you might wish to know.

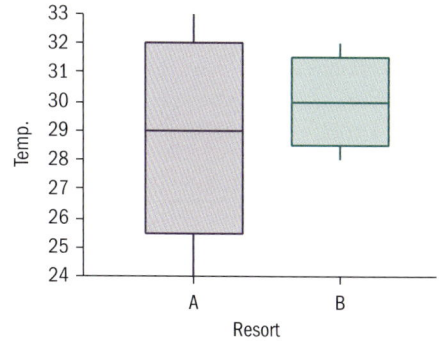

2. The two box-and-whisker below relate to the salaries in a small college in the USA.

    a) The first shows the salaries paid during the 1991–1992 academic year, separately for males and females. Is there any difference in the treatment of men and women in this college?

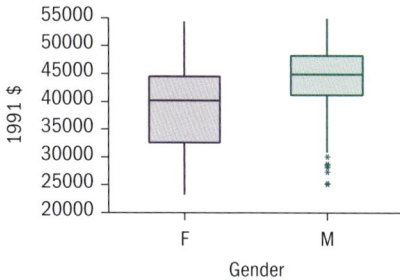

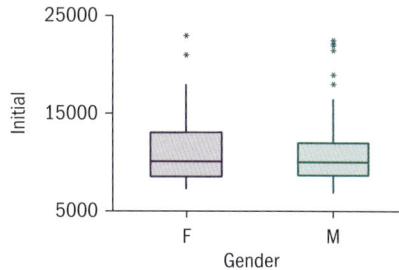

> This is a real data set used in fighting discrimination which illustrates why simple data sets in complex contexts can be difficult to interpret meaningfully.

    b) The second shows the staff members' initial salary, in whatever year they joined the staff, again shown separately for males and females. Does the information contained in this diagram alter any views you formed from the first dataset? If so, explain how.

3. The quartiles of the number of employees in random samples of firms in 2010 and in 2015 are given in a table below, together with any outliers which are more than $1.5 \times$ the interquartile range beyond the quartile.

	Min	LQ	Median	UQ	Max	Outliers
2010	1	6	18	42	145	106, 131, 145
2015	1	3	13	35	160	95, 160

    a) Construct a box-and-whisker plot for each year.

    b) Compare and comment on the two distributions.

4. The table shows the serving sizes the US Food and Drug Administration use for 15 foods in each of the nutrition factsheets they issue for fruit, vegetables and sea food.

Fruit	242	30	126	134	126	134	148	154	147	166	112	161	147	140	280
Vegetables	93	148	148	78	99	110	99	84	89	85	84	148	148	90	148
Seafood	84	84	84	84	84	84	84	84	84	84	84	84	84	84	84

a) Find the quartiles for each of the three types of food.

b) Why would you not draw a boxplot for the serving sizes of seafood?

c) Using a single axis, construct box-and-whisker plots for the serving sizes of fruit and vegetables in the table.

5. The diagram shows the scores registered for the 100 metres, the long jump and the 1500 metres in the personal bests for the top 75 decathletes up to June 2014. Compare the performances in the 3 events.

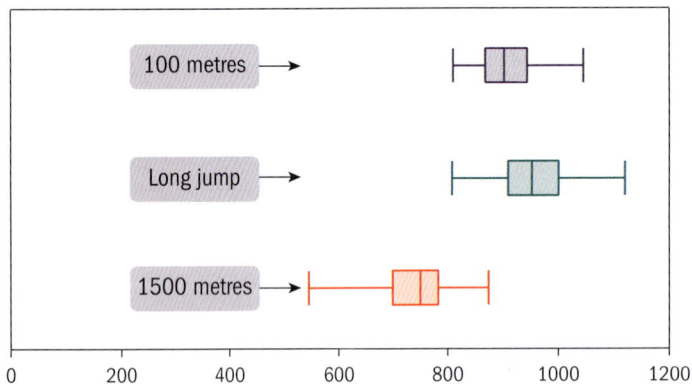

3.3 Histograms

In a histogram, the area of each bar is proportional to the frequency in the interval.

When the intervals are not of equal width the height of the bar is the **frequency density** and it must be scaled.

● $\left[\text{Frequency density} = \dfrac{\text{frequency}}{\text{interval width}}\right].$

Example 2

The heights of the children in a school are measured correct to the nearest centimetre and are summarised in the table:

Height (cm)	120–129	130–139	140–144	145–149	150–154	155–159	160–169	170–179
Frequency	60	80	50	93	77	67	72	54

Draw a histogram to represent this data.

The interval end-points are

119.5, 129.5, 135.4, 139.5, 144.5, 149.5, 154.5, 159.5, 169.5 and 179.5

so the widths of the intervals are 10, 10, 5, 5, 5, 5, 10 and 10.

> It is common to have wider intervals when there is less data to collect.
>
> This 'smoothes' the data and makes it less likely you will over-interpret or over-emphasise rare data observations.

The heights of the bars will then be

$\frac{60}{10} = 6; \frac{80}{10} = 8; \frac{50}{5} = 10; \frac{93}{5} = 18.6; \frac{77}{5} = 15.4; \frac{67}{5} = 13.4; \frac{72}{10} = 7.2$ and $\frac{54}{10} = 5.4$

and the histogram looks like:

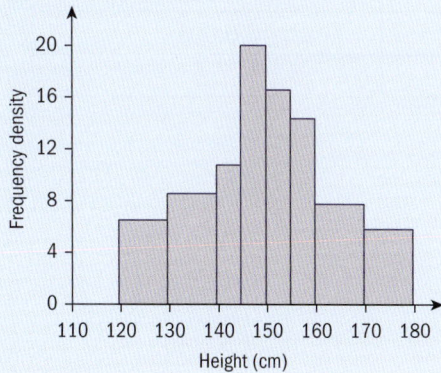

> The horizontal axis should be drawn as a linear scale showing the variable.

Relative frequency histograms*

> **Note:** This section is not required for the Cambridge 9709 syllabus, but it may aid in your understanding.

> Continuous probability distributions are defined like this – see page 153.

In Chapter 8 you will meet the Normal distribution which is an example of a probability distribution.

In a **relative frequency histogram**, the total area is defined to be 1 square unit, and then the area of the histogram between any two values will represent the proportion of data which lies between those two values.

To calculate the heights of the bars for a relative frequency histogram, you can divide the frequency densities by the total frequency – this will automatically scale the total area to be 1.

The earlier example looks like the distribution on the right.

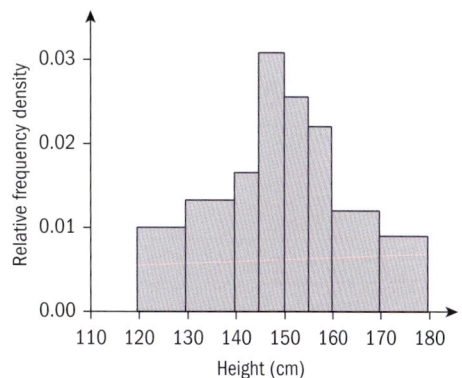

Example 3

The ages of members of a cricket society are summarised in the table below.

Age (years)	25–34	35–44	45–49	50–54	54–59	60–74
Frequency	6	9	13	9	7	6

a) Draw a relative frequency histogram to display this information.

b) Calculate an estimate of the number of members aged between 40 and 52, inclusive.

Since these are ages, the ends of the first interval are at 25 and 35 etc. A table is the best way to show the calculations for either sort of histogram

Ages	Frequency	Interval width	Frequency density	Relative frequency density
25–34	6	10	0.6	0.012
35–44	9	10	0.9	0.018
45–49	13	5	2.6	0.052
50–54	9	5	1.8	0.036
55–59	7	5	1.4	0.028
60–74	6	15	0.4	0.008
Total	50			

To estimate the number of members between 40 and 52, it includes half the time of the 35 – 44 interval, all of 45 – 49 and $\frac{3}{5}$ of the 50 – 54 interval.

Taking the same proportions of the numbers in the intervals gives
$0.5 \times 9 + 13 + 0.6 \times 9 = 22.9$ as the estimate.

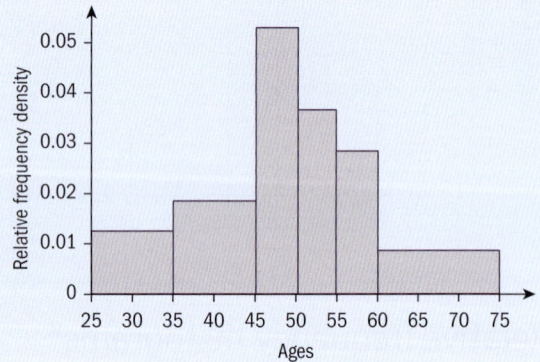

Exercise 3.3

1. Draw a histogram of the following data:

Length (cm)	20–50	50–60	60–65	65–70	70–80	80–90	90–110
Number of plants	27	18	16	15	22	14	14

2. The table gives the daily protein intake of a sample of 130 people in a country.

Protein (grams)	0–15	–25	–30	–35	–40	–45	–50	–55	–65
Number of people	18	14	11	16	21	18	15	11	6

a) Draw a histogram for these data

b) Calculate estimates of the mean and standard deviation of the daily protein intake of the people in the country.

c) Corresponding data were collected in another country. The mean of that set was 45.1 grams and the standard deviation was 5.9 grams. What does this tell you about the diets of the two countries?

3. A question on a survey asked the students how long their journeys to school had taken on a particular day. A summary of the results is shown below.

Time (minutes, correct to the nearest minute)	Number of children
1–15	115
16–25	46
26–35	36
36–55	22
56–80	14

Illustrate these data using an appropriate diagram.

4. The table shows information about the salaries paid to employees in a company.

Salary	Frequency
$0 < x \le \$10\,000$	7
$\$10\,000 < x \le \$15\,000$	82
$\$15\,000 < x \le \$20\,000$	45
$\$20\,000 < x \le \$25\,000$	24
$\$25\,000 < x \le \$30\,000$	13
$\$30\,000 < x \le \$50\,000$	4

Draw a histogram to represent these data.

5. The length of time students take to complete a mathematical puzzle is summarised in the table.

a) Draw a relative frequency histogram to represent these data.

b) Calculate an estimate of the number of students who took more than 10 minutes, but not more than 15 minutes, to complete the puzzle.

Time, t (minutes)	Number of pupils
$0 < t \le 2$	6
$2 < t \le 5$	8
$5 < t \le 8$	15
$8 < t \le 12$	14
$12 < t \le 20$	7

3.4 Cumulative frequency graphs

When data is given in a grouped frequency table, you can use **interpolation** to estimate the value of quartiles.

Note: You will normally be asked to do this by drawing a cumulative frequency graph, but this example will show you what is happening when you use the CF graph.

Example 4

A question on a survey asks students how long their journey to school took on a particular day. A summary of the results is shown below.

Time (minutes, correct to the nearest minute)	Number of children
1–15	115
16–25	46
26–35	36
36–55	22
56–80	14

Calculate estimates of **a)** the median and **b)** the upper quartile for this data.

Rewrite the table with cumulative frequencies:

Time (minutes, correct to the nearest minute)	Number of students	Cumulative frequency
1–15	115	115
16–25	46	161
26–35	36	197
36–55	22	219
56–80	14	233

a) n = 133, and the median is the 117th value 117 is between 115 and 161, so the upper quartile lies in the 16 – 25 interval.

117 – 115 = 2 and there are 46 students in the interval, so the estimate will be $\frac{2}{46}$ of the way through the interval.

The interval starts at 15.5 and is of width 10 minutes so the upper quartile is

$$15.5 + \frac{2}{46} \times 10 = 15.9347... = 15.9 \text{ minutes.}$$

▶ Continued on the next page

b) $n = 233$, and the upper quartile is the 175th value ($233 \times \frac{3}{4} = 174.75$)

175 is between 161 and 197, so the upper quartile lies in the 26–35 interval.

$175 - 161 = 14$ and there are 36 students in the interval,

so the estimate will be $\frac{14}{36}$ of the way through the interval.

The interval starts at 25.5 and is of width 10 minutes so the upper quartile is

$25.5 + \frac{14}{36} \times 10 = 29.3888... = 29.4$ minutes.

You may prefer to use a formula for interpolation:

• Estimated value of quartile =

lower class boundary $+ \left(\dfrac{Q \text{ - cum. freq. at start of interval}}{\text{interval frequency}} \right) \times$ interval width

where Q is the position of the quartile required.

> An alternative approach is to plot the cumulative frequencies on a graph, using the interval end-points as the x-coordinate and the cumulative frequency as the y-coordinate.

If you then join these points with straight lines, or by drawing the smoothest curve through them that you can, you can use the graph to estimate the times for any number of students you are interested in, including the quartiles.

The points to be plotted are (0.5, 0); (15.5, 115); (25.5, 161); (35.5, 197); (55.5, 219); (80.5, 233) – remember to look carefully at what the interval end-points will be, it is normally easy to get them right provided you do think specifically about how the data was recorded and any rounding that took place.

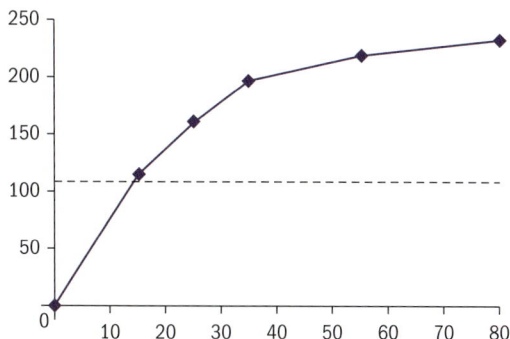

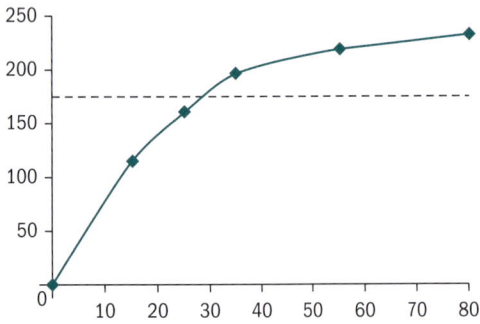

We can see that, for this set of data, the two graphs are not very different. Either approach is acceptable for this examination so choose one, stick with it, and draw it clearly.

- The straight line graph is effectively spacing all the data values within an interval equally, and changes the rate of occurrence at the end of each interval.

- The interval end-points are often arbitrary cutoffs and there is not actually any change in the rate of occurrence there – the changes are more gradual and occur continuously.

- The 'best smooth curve' is trying to produce the best fit – it lets the intervals before and after influence what we think will happen in an interval – for example if there is 40 in an interval and then 75 in the one above but only 6 in the one below, we would expect more of the 40 to occur towards the top end of that middle interval.

Exercise 3.4

1. A group of students recorded their pulse rates when they were in a relaxed restful state.

53	54	55	55	57	58	59	59	59	62	62	62
62	62	63	63	63	64	64	65	65	67	67	67
68	68	69	69	70	71	71	73	75	79	80	83

a) Calculate the median and the quartiles of these pulse rates

b) Construct a grouped frequency table for the students' pulses, using intervals 51–55; 56–60; 61–65; 66–70; 71–75; 76–80; 81–85

c) Calculate estimates of the median and the quartiles of these pulse rates using the table in part (b) using interpolation

> We normally only have the grouped frequency data, but these examples will give us a feel for how good the estimates are likely to be when we are only working with the summarised data.

2. A group of 45 women go on a skiing holiday.

Their ages are listed below.

17	17	17	19	19	19	20	22	22	22	23	23	24	25	25
25	26	26	26	26	27	27	28	28	28	29	29	29	30	32
32	32	34	34	35	35	35	35	36	37	38	38	44	45	48

    a) Calculate the median and quartiles of their ages

    b) Construct a grouped frequency table for the womens' ages, using intervals 17–20; 21–24; 25–28; 29–32, 33–48

    c) Draw a cumulative frequency graph for the data table in part (**b**)

    d) Use your cumulative frequency graph to estimate the median and quartiles of their ages.

3. The table shows information about the salaries paid to employees in a company.

Salary	Frequency
$0 < x \leq \$10\,000$	24
$\$10\,000 < x \leq \$15\,000$	127
$\$15\,000 < x \leq \$20\,000$	45
$\$20\,000 < x \leq \$25\,000$	24
$\$25\,000 < x \leq \$30\,000$	13
$\$30\,000 < x \leq \$50\,000$	4

    a) Draw a cumulative frequency graph for the salaries data

    b) Use your graph to estimate the median and quartiles.

    c) Estimate the proportion of the employees who earn more than twice the median salary?

4. The length of time pupils take to complete a mathematical puzzle is summarised in the table.

Time, t (minutes)	Number of pupils
$0 < t \leq 2$	6
$2 < t \leq 5$	8
$5 < t \leq 8$	15
$8 < t \leq 12$	14
$12 < t \leq 20$	7

    a) Draw a cumulative frequency graph to represent these data.

Calculate estimates of

    b) the median and

    c) the interquartile range of the times taken.

5. The arrival times of 204 trains were noted and the number of minutes, t, that each train was late was recorded. The results are summarised in the table.

Number of minutes late (t)	$-2 \leq t < 0$	$0 \leq t < 2$	$2 \leq t < 4$	$4 \leq t < 6$	$6 \leq t < 10$
Number of trains	43	51	69	22	19

i) Explain what $-2 \leq t < 0$ means about the arrival times of trains. [1]

ii) Draw a cumulative frequency graph, and from it estimate the median and the interquartile range of the number of minutes late of these trains. [7]

Cambridge International AS and A level Mathematics 9709,
Q5 October/November 2007 Paper 6

6. During January the numbers of people entering a store during the first hour after opening were as follows.

Time after opening, x minutes	Frequency	Cumulative frequency
$0 < x \leq 10$	210	210
$10 < x \leq 20$	134	344
$20 < x \leq 30$	78	422
$30 < x \leq 40$	72	a
$40 < x \leq 60$	b	540

i) Find the values of a and b. [2]

ii) Draw a cumulative frequency graph to represent this information. Take a scale of 2 cm for 10 minutes on the horizontal axies and 2 cm for 50 people on the vertical axis. [4]

iii) Use your graph to estimate the median time after opening that people entered the store. [2]

iv) Calculate estimates of the mean, m minutes, and standard deviation, s minutes, of the time after opening the people entered the store. [4]

v) Use your graph to estimate the number of people entering the store between $\left(m - \frac{1}{2}s\right)$ and $\left(m + \frac{1}{2}s\right)$ minutes after opening. [2]

Cambridge International AS and A Level Mathematics 9709,
Q6 May/June 2009 Paper 6

3.5 Skewness

> **Skewness** is the term used to describe the lack of symmetry in a distribution.

Histograms and box-and-whisker can show skewness quite well, particularly where data is not symmetric about the median.

A **positively skewed** distribution has:
- o   A long tail to the right
- o $Q_3 - Q_2 > Q_2 - Q_1$
- o   mode < median < mean

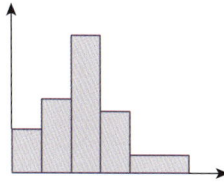

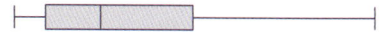

A **negatively skewed** distribution has:
- o   A long tail to the left
- o $Q_3 - Q_2 < Q_2 - Q_1$
- o   mean < median < mode

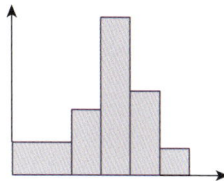

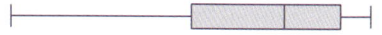

A **symmetric** distribution has:
- o   Equal length tails
- o $Q_3 - Q_2 \sim Q_2 - Q_1$

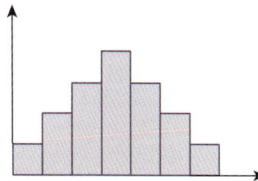

There are some common measures of skewness:

These measures quantify skewness as well as giving the sign (positive or negative).

$$\frac{3 \times (\text{mean-median})}{\text{standard deviation}} \quad \text{and} \quad \frac{\text{mean-mode}}{\text{standard deviation}}$$

The quartile skewness coefficient is defined by

$$\frac{(Q_3 - Q_2) - (Q_2 - Q_1)}{Q_3 - Q_1} = \frac{Q_3 - 2Q_2 + Q_1}{Q_3 - Q_1}.$$

All these measures allow comparisons to be made between different distributions.

> You do not need to learn these measures. Any question which includes skewness formally will give a formula to calculate a coefficient of skewness, but you do need to be able to describe skewness informally.

Example 5

Here are the summary statistics for some data:

Mean = 10.7,

standard deviation = 9.3,

lower quartile = 3.6,

median = 8.6,

upper quartile = 15.4 and

mode = 2.7

The boxplot is shown on the right.

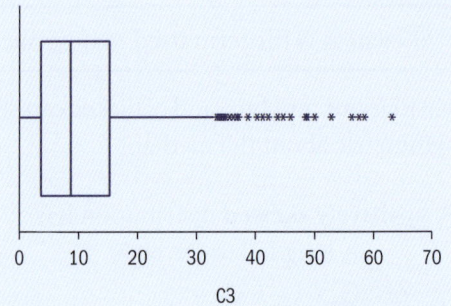

a) Describe the skewness of this distribution, giving two reasons.

b) Calculate the skewness coefficient $\dfrac{3\times(\text{mean-median})}{\text{standard deviation}}$ and state what it means.

· ·

a) There is a positive skew since

   i) the boxplot shows a long tail to the right

   ii) mode < median < mean

b) $\dfrac{3\times(\text{mean-median})}{\text{standard deviation}} = \dfrac{3\times(10.7-8.6)}{9.3} = \dfrac{6.3}{9.3} = 0.68$

   The positive value indicates that the skew is positive.

Note in the last example you could have worked out other skewness coefficients:

$\dfrac{\text{mean-mode}}{\text{standard deviation}} = \dfrac{10.7-2.7}{9.3} = \dfrac{8}{9.3} = 0.86$

$\dfrac{(Q_3-Q_2)-(Q_2-Q_1)}{Q_3-Q_1} = \dfrac{Q_3-2Q_2+Q_1}{Q_3-Q_1} = \dfrac{15.4-2\times8.6+3.6}{15.4-3.6} = \dfrac{1.8}{11.8} = 0.15$

The numerical value of each skewness coefficient is different. To compare distributions it is therefore important that you use the same measure.

Outliers

Outliers are extreme values, and they are common in skewed distributions. An outlier is often regarded as being:

You met outliers briefly on page 40.

- Any value which is more than $1.5 \times \text{IQR}$ above the Upper Quartile or below the Lower Quartile

  $x - Q_3 > 1.5\times(Q_3 - Q_1)$

  $Q_1 - x > 1.5\times(Q_3 - Q_1)$

  Or...

- Any value which is more than 2 standard deviations above or below the mean $\left|\dfrac{x-\bar{x}}{\sigma}\right| > 2$

Exercise 3.5*

1. Summary statistics are given for three groups in the table below.

	Group A	Group B	Group C
Mean	35.7	89.2	54.0
Standard deviation	6.4	11.3	7.2
Median	32.1	95.3	56.5
Lower quartile	28.3	81.6	47.2
Upper quartile	41.2	102.5	62.7
Mode	29.1	96.1	55.8

a) Calculate the coefficient of skewness $\dfrac{3\times(\text{mean-median})}{\text{standard deviation}}$ for each of the groups.

b) Which of the three distributions is most skewed by this measure?

2. Summary statistics are given for three groups in the table below.

	Group A	Group B	Group C
Mean	84.1	79.3	56.1
Standard deviation	6.2	17.2	8.3
Median	85.1	84.2	52.1
Lower quartile	79.3	71.1	41.0
Upper quartile	90.1	102.7	59.3
Mode	86.2	85.3	53.1

a) Calculate the coefficient of skewness $\dfrac{Q_3 - 2Q_2 + Q_1}{Q_3 - Q_1}$ for each of the groups.

b) Which of the three distributions is most skewed by this measure?

c) Does the coefficient of skewness $\dfrac{\text{mean-mode}}{\text{standard deviation}}$ give the same order of skewness for these groups?

3. The ages of applicants for mortgages is recorded by an estate agency. The results are shown below.

25, 29, 27, 32, 45, 34, 26, 28, 30, 42, 26, 51, 29, 27, 33, 27

a) Calculate the mean of these data.

b) Draw a stem-and-leaf diagram to represent these data.

c) Find the median and the quartiles of these data.

An outlier is an observation that falls either $1.5 \times$ (interquartile range) above the upper quartile or $1.5 \times$ (interquartile range) below the lower quartile.

d) Determine whether or not any items of data are outliers.

e) On graph paper draw a boxplot to represent these data. Show your scale clearly.

f) Comment on the skewness of the distribution of ages of applicant for mortgages. Justify your answer.

4. At a metro station, a regular passenger times (in seconds) how long he has to wait for a train to arrive once he gets to the platform. These data are listed below.

| 87 | 42 | 0 | 62 | 124 | 0 | 58 | 37 | 74 | 94 |
| 182 | 23 | 17 | 62 | 29 | 17 | 82 | 54 | 0 | 45 |

a) Find the median and inter-quartile range of the waiting times.

An outlier is an observation that falls either $1.5 \times$ (inter-quartile range) above the upper quartile or $1.5 \times$ (inter-quartile range) below the lower quartile.

b) Draw a boxplot to represent these data, clearly indicating any outliers.

c) Comment on the skewness of these data. Justify your answer.

d) Explain how a zero waiting time occurs.

3.6 Comparing distributions

There are various measures of average that you can use to compare distributions, and you are not expected to calculate all of them.

The **mean** uses all data values, but can be distorted by outliers.

The **median** is the middle value, and is less influenced by outliers.

The **modal class** has the highest frequency density – in a histogram, it is the interval with the tallest block

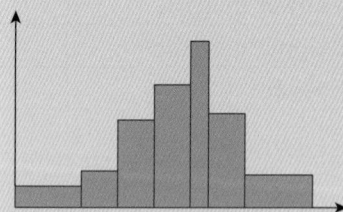

Similarly there are various measures of spread as well.

The **standard deviation** is used with the mean. It can therefore be distorted by outliers.

The **inter-quartile range** is used with the median. It concerns the middle 50%, and is unaffected by outliers.

The **range** is the difference between the highest and lowest values, so it is badly affected by extreme values.

- If you are comparing distributions,
  - make the comparison in context
  - make reference to both the average values and the spread.

It is easier to make comparisons between two distributions than it is to describe one set of data from a graph. Histograms, box-and-whisker, and stem-and-leaf diagrams all show average values, spread and whether the distribution is reasonably symmetric.

The following data are marks on a test out of 30 for two classes, A and B.

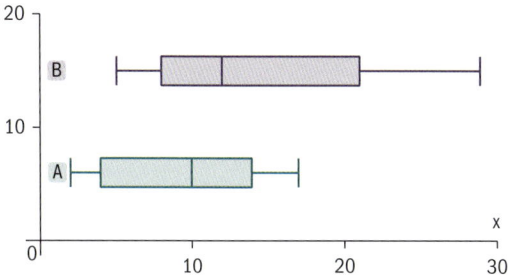

On average, class B have done better than class A. There is a much greater spread of marks in class B than in class A.
The other striking feature here is that the top half of the marks in class B are very spread out in comparison with the bottom half of B and both halves of A.

Example 6

The times taken by pupils from three schools (A, B, C) to complete a mathematical challenge are summarised in the box-and-whisker below. Compare the performance of the pupils in the three schools.

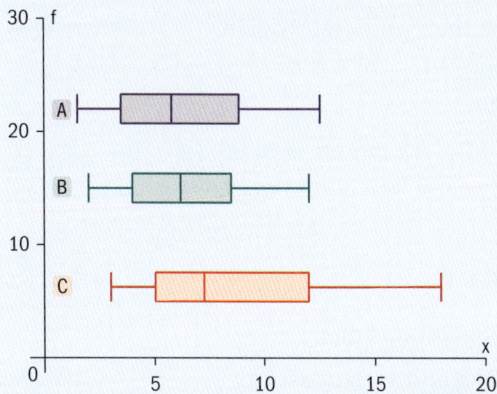

Pupils in schools A and B generally did better than school C – they completed the puzzle quicker and there was less variability in the times they took. A had the quickest pupils but also had pupils slower than any of B's.

> You could say a little more, but be careful not to go into too much detail.

In cumulative frequency graphs, the shape of the distribution is much less clear, but it is worth remembering that the gradient of the CF graph represents the rate at which observations occur – which is the height of the histogram at that point in the distribution.

In comparative bar charts, make sure you talk about comparisons between each of the variables – so if you have data on boys and girls of different ages, talk about both – even if there is no difference across one variable.

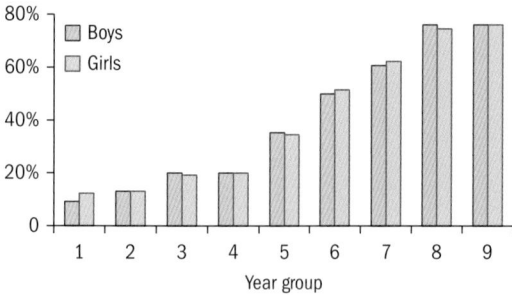

The proportion of boys and girls owning mobile phones is very similar, but older pupils are more likely to own a mobile phone.

Exercise 3.6

1. A group of students took their pulse rates at the end of a mathematics lesson. The mean was 64.2 and the standard deviation was 6.5.

   At the end of the next lesson, which was PE, they took their pulse rates again. The mean was now 71.6 and the standard deviation was 9.1.
   Compare the pulse rates before and after the PE lesson.

2. A new diet is claimed to increase the speed of weight loss. A random sample of 12 volunteers followed the diet and recorded their weight loss in one week. The results were (in kilograms):

   1.3   1.1   0.5   1.3   1.5   1.4   0.8   1.2   1.0   0.8   1.2   1.1

   a) Calculate the mean and variance of the weight loss with this diet.

   A traditional diet has a weight loss with a mean of 0.8 kg and variance 0.1 kg^2
   b) Compare the weight loss from the two diets.

3. The box-and-whisker show the lengths (in cm) of a certain type of plant found in two gardens. Compare the plants in gardens A and B.

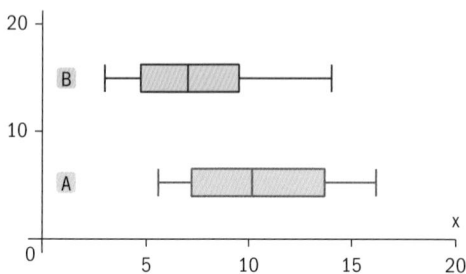

Summary exercise 3

1. Air traffic control at an airport record the delays in arrival times of flights into the airport. On one day, 40 % of all flights had no delay, the longest delay was 44 minutes and half of all flights had delays of no more than 4 minutes. A quarter of all delays were at least 18 minutes, but only one was more than 25 minutes.

   An outlier is an observation that falls either $1.5 \times$ (interquartile range) above the upper quartile or $1.5 \times$ (interquartile range) below the lower quartile.

   a) On graph paper, draw a boxplot to represent these data.

   b) Comment on the distribution of delays. Justify your answer.

   The box plot below summarises the delays at the same airport on another day.

   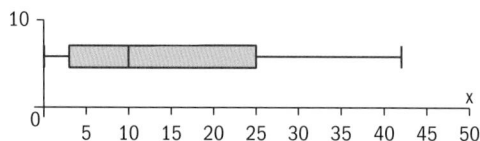

   c) Compare the delays to flights arriving at the airport on the two days.

2. An airport bus service runs from the city centre, and from the airport, every 10 minutes during the day. The number of passengers on a random sample of journeys from the airport is shown below.

   11, 3, 14, 34, 1, 6, 12, 15, 8, 4, 28, 7, 3, 0, 8, 12.

   Draw a stem-and-leaf diagram to represent these data.

EXAM-STYLE QUESTION

3. A regular airline passenger keeps a record of the amount of time which passes after the plane lands before she has collected her bags and exited the terminal. The times, to the nearest minute, for all the journeys she has made in the past six months are summarised in the table below.

Time to exit terminal	Number of flights
5–6	12
7–9	15
10–14	12
15–19	8
20–29	11
30–49	14

   a) Give a reason to support the use of a histogram to represent these data.

   b) Write down the upper class boundary and the lower class boundary of the class 10–14.

   c) On graph paper, draw a histogram to represent these data.

   d) Calculate estimates of the mean and standard deviation of the time to exit the terminal.

   On international flights she had to go through passport control and immigration as well as collecting baggage before exiting the terminal. The shortest time she took to exit on any of the 12 international flights was 25 minutes.

   e) State, with reasons, what effect excluding the international flights may have on the size of:

      i) the mean

4. A random sample of 97 people who own mobile phones was used to collect data on the amount of time they spent per day on their phones. The results are displayed in the table below.

Time spent per day (t minutes)	$0 \le t < 5$	$5 \le t < 10$	$10 \le t < 20$	$20 \le t < 30$	$30 \le t < 40$	$40 \le t < 70$
Number of people	11	20	32	18	10	6

i) Calculate estimates of the mean and standard deviation of the time spent per day on these mobile phones. [5]

ii) On graph paper, draw a fully labelled histogram to represent the data. [4]

Cambridge International AS and A level Mathematics 9709, Paper 6 Q7 May/June 2003

5. The lengths of cars travelling can a car ferry are noted. The data are summarised in the following table.

Length of car (x metres)	Frequency	Frequency density
$2.80 \le x < 3.00$	17	85
$3.00 \le x < 3.10$	24	240
$3.10 \le x < 3.20$	19	190
$3.20 \le x < 3.00$	8	a

i)   Find the value of a. [1]

ii) Draw a histogram on graph paper to represent the data. [3]

iii) Find the probability that a randomly chosen car on the ferry is less than 3.20 m in length. [2]

Cambridge International AS and A level Mathematics 9709, Paper 6 Q2 October/November 2004

6. In a recent survey, 640 people were asked about the length of time each week that they spent watching television. The median time was found to be 20 hours, and the lower and upper quartiles were 15 hours and 35 hours respectively. The least amount of time that anyone spent was 3 hours, and the greatest amount was 60 hours.

i) On graph paper, show these results using a fully labelled cumulative frequency graph. [3]

ii) Use your graph to estimate how many people watched more than 50 hours of television each week. [2]

Cambridge International AS and A level Mathematics 9709, Paper 6 Q2 May/June 2004

7. The following back-to-back stem-and-leaf diagram shows the cholesterol count for a group of 45 people who exercise daily and for another group of 63 who do not exercise. The figures in brackets show the number of people corresponding to each set of leaves.

	People who exercise			People who do not exercise	
(9)	9 8 7 6 4 3 2 2 1	3	1 5 7 7		(4)
(12)	9 8 8 8 7 6 6 5 3 3 2 2	4	2 3 4 4 5 8		(6)
(9)	8 7 7 7 6 5 3 3 1	5	1 2 2 2 3 4 4 5 6 7 8 8 9		(13)
(7)	6 6 6 6 4 3 2	6	1 2 3 3 3 4 5 5 5 7 7 8 9 9		(14)
(3)	8 4 1	7	2 4 5 5 6 6 7 8 8		(9)
(4)	9 5 5 2	8	1 3 3 4 6 7 9 9 9		(9)
(1)	4	9	1 4 5 5 8		(5)
(0)		10	3 3 6		(3)

Key: 2|8|1 represents a cholesterol count of 8.2 in the group who exercise and 8.1 in the group who do not exercise.

i) Give one useful feature of a stem-and-leaf diagram. [1]

ii) Find the median and the quartiles of the cholesterol count for the group who do not exercise. [3]

You are given that the lower quartile, median and upper quartile of the cholesterol count for the group who exercise are 4.25, 5.3 and 6.6 respectively.

iii) On a single diagram on graph paper, draw two box-and-whisker plots to illustrate the data. [4]

Cambridge International AS and A level Mathematics 9709, Paper 6 Q4 May/June 2005

8. Each father in a random sample of fathers was asked how old he was when his first child was born. The following histogram represents the information.

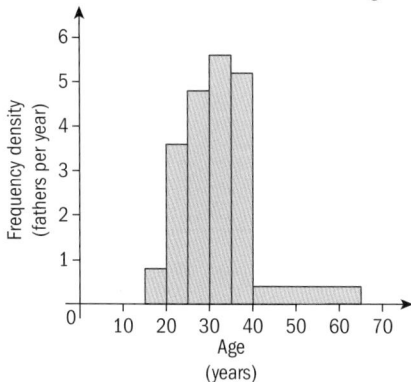

i) What is the model age group? [1]

ii) How many fathers were between 25 and 30 years old when their first child was born? [2]

iii) How many fathers were in the sample? [2]

iv) Find the probability that a father, chosen at random from the group, was between 25 and 30 years old when his first child was born, give that he was older than 25 years. [2]

Cambridge International AS and A level Mathematics 9709, Paper 6 Q5 May/June 2006

9. The following table has some statistical data for 18 countries relating to 2012, published by UNICEF.

UNICEF statistics	Infant (under 1) mortality rate (deaths per thousand births), 2012	Total population (thousands) 2012	GNI per capita (US$) 2012	Life expectancy at birth (years) 2012	Total adult literacy rate (%) 2008–2012*
Bangladesh	33	154695.4	840	70.3	57.7
Brunei Darussalam	7	412.2	d	78.4	95.4
China	12	1377064.9	5740	75.2	95.1
Egypt	18	80721.9	3000	70.9	73.9
India	44	1236686.7	1530	66.2	62.8
Indonesia	26	246864.2	3420	70.6	92.8
Malaysia	7	29239.9	9800	74.8	93.1
Mauritius	13	1239.6	8570	73.5	88.8
Nepal	34	27474.4	700	68	57.4
New Zealand	5	4459.9	30620	81	–
Pakistan	69	179160.1	1260	66.4	54.9
Saudi Arabia	7	28287.9	18030	75.3	87.2
South Africa	33	52385.9	7610	56.3	93
Trinidad and Tobago	18	1337.4	14400	69.8	98.8
United Arab Emirates	7	9205.7	36040	76.7	90
United Kingdom	4	62783.1	38250	80.4	–
United States	6	317505.3	50120	78.8	–
Zimbabwe	56	13724.3	680	58.1	83.6

a) Construct a stem-and-leaf diagram for the infant mortality rates for these countries.

b) Construct a stem-and-leaf diagram for the total adult literacy rates (rounded to the nearest percent) for these countries.

c) Construct a box-and-whisker plot for the life expectancy at birth for these countries. Describe the skewness of the data.

d) Construct a box-and-whisker plot for the GNI (Gross National Income) per capita for these countries. Describe the skewness of the data.

e) Construct a box-and-whisker plot for the total population of these countries. Describe the skewness of the data.

10. In a particular week, a clinic nurse treats a number of patients. The length of time, to the nearest minute, for each patient's treatment is summarised in the table below.

Time (minutes)	3–6	7–8	9–10	11–12	13–15	16–20
Number of patients	15	12	17	15	16	15

Draw a histogram to illustrate these data.

11. The table below shows the points scored for the first 3 disciplines of the top twenty personal best decathlon performances of all time. Construct box-and-whisker plots of the data for the 100 metres, the long jump and the shot put in one diagram and compare the points scored in the three disciplines.

Athlete	100 m	Long	Shot
Ashton Eaton	1044	1120	741
Roman Šebrle	942	1089	810
Tomaš Dvorak	966	1035	899
Dan O'Brien	992	1081	894
Daley Thompson	989	1063	834
Jürgen Hingsen	929	1000	877
Bryan Clay	1001	908	800
Erki Nool	952	967	784
Uwe Freimuth	847	1007	870
Trey Hardee	987	1017	810
Tom Pappas	910	1050	869
Siegfried Wentz	885	932	811
Eduard Hämäläinen	975	876	854
Dmitri Karpov	975	1012	847
Aleksandr Apaichev	870	952	851
Frank Busemann	952	1079	704
Dave Johnson	870	918	766
Grigori Degtyaryov	890	915	853
Chris Huffins	1020	1000	816
Torsten Voss	931	1030	788

The full data set can be found at the end of the book.

Chapter summary

- Stem-and-leaf diagrams give the shape of a distribution in the same way as a histogram with equal intervals does, but also keep the detail available.

- Back-to-back stem-and-leaf diagrams allow distributions to be compared citing position of centre and spread in context.

- Box-and-whisker plots show only five 'points' from the distribution, so they are good for an overview in comparing distributions but very short of detail.

- In a histogram, the area of each bar is proportional to the frequency in the interval. When the intervals are not of equal width the height of the bar is the **frequency density** and it must be scaled.

- A cumulative frequency graph plots the interval end-points as the x coordinate and the cumulative frequency as the y-coordinate. These points can be joined by straight lines or by a smooth curve through and used to estimate medians and quartiles.

- Skewness is the term used to describe the lack of symmetry in a distribution.
  - A positively skewed distribution has a long tail to the right.
  - A negatively skewed distributed has a long tail to the left.
  - A symmetric distribution has equal length tails.

- If you are comparing distributions,
  - make the comparison in context
  - make reference to both the average values and the spread.

Review exercise A

1. A cricketer keeps a record of his scores in a club league during 2014.

   They were 0, 25, 0, 122, 7, 36, 54, 0, 3, 97, 45, 0, 107

   Calculate
   a) the mode,
   b) the median and
   c) the mean of his scores.

2. A theme park records the number of tickets being bought together by individuals or groups during a one hour period.

Number of people in the group	Frequency
1	12
2	9
3	8
4	12
5	7
6	5
7	3
8	1

   a) Calculate i) the mode, ii) the median and ii) the mean size of group (counting an individual as a group).
   b) How many tickets were bought during the hour?

3. The table show information about the heights of a number of plants of a particular species.

Length (cm)	20–50	50–60	60–65	65–70	70–80	80–90	90–110
Number of plants	15	22	16	15	21	16	11

   a) Calculate an estimate of the mean and variance of the height of these plants.
   b) Draw a histogram to show this data.

4. $\sum x = 125$ $\sum x^2 = 963.2$ $n = 20$ Find the mean and variance of X.

5. $\sum (x - 35) = 4.8$ $\sum (x - 35)^2 = 35.2$ $n = 8$ Find the mean and standard deviation of X.

6. $\sum (x - 25) = -20$ $\sum (x - 25)^2 = 231.2$ $n = 24$ Find $\sum x$, $\sum x^2$

7. A set of observations $\{x\}$ are coded using $X = \dfrac{x - 87.5}{10}$.

   Given that $\overline{X} = 5.9$ $\mathrm{Var}(X) = 16.3$, calculate the mean and variance of the original set of observations.

8. The temperatures (in °C) at a resort are measured at the same time on eight successive Mondays:

   15.6, 18.2, 17.3, 19.2, 20.3, 23.4, 22.1, 20.7
   a) Calculate the mean of these temperatures (in °C).
   b) Convert each temperature into °F by the formula $F = 32 + 1.8°C$
   c) Calculate the mean of these temperatures (in °F).
   d) Check that the mean temperature in °C converts to the same mean in °F.

9. The marks for a class of 20 students has $\sum x = 1313$, and $\sum x^2 = 88\,261$.
   a) Calculate the mean and variance of the marks for this class.

   A second class of 15 students on the same examination had a mean of 63.8 and standard deviation of 5.58 marks.

   b) Calculate $\sum x^2$ for the second class.
   c) Calculate the mean mark of all the students in the two classes.
   d) Comment on the performance of the two classes on the examination.

10. The marks on a mathematics examination were 82, 51, 37, 64, 72.
    a) Calculate the mean and standard deviation of these marks.

    For comparisons with other subjects, the marks are to be scaled so that they have a mean of 50 and a standard deviation of 10. This is to be done by using the transformation $y = \dfrac{(x-a)}{b}$ where X is the original mark and Y is the transformed mark.
    b) Calculate the values of a and b.

11. The table below shows the points scored for the first 3 disciplines of the top twenty personal best decathlon performances of all time. Construct box-and-whisker plots of the data for the Pole vault, the javelin and the 1500 metres in one diagram and compare the points scored in the three disciplines.

Athlete	pole	javelin	1500 m
Ashton Eaton	1004	721	850
Roman Šebrle	849	892	798
Tomaš Dvorak	880	925	698
Dan O'Brien	910	777	667
Daley Thompson	910	817	712
Jürgen Hingsen	880	736	813
Bryan Clay	910	898	613
Erki Nool	1035	844	747
Uwe Freimuth	957	926	776
Trey Hardee	972	859	625
Tom Pappas	972	749	630
Siegfried Wentz	849	900	778
Eduard Hämäläinen	880	743	712
Dmitri Karpov	790	671	692
Aleksandr Apaichev	880	924	768
Frank Busemann	849	842	735
Dave Johnson	998	843	749
Grigori Degtyaryov	880	845	791
Chris Huffins	790	762	563
Torsten Voss	941	708	772

12. An airport bus service runs from the city centre, and from the airport, every 10 minutes during the day. The number of passengers on a random sample of journeys from the airport is shown below.

21, 13, 17, 24, 11, 6, 22, 12, 6, 3, 25, 17, 13, 0, 5, 12, 16, 7, 22, 3, 18, 31

Draw a stem-and-leaf diagram to represent these data.

13. The table shows information about the salaries paid to employees in a company.

Salary	Frequency
$0 < x \leq \$10\,000$	9
$\$10\,000 < x \leq \$15\,000$	65
$\$15\,000 < x \leq \$20\,000$	72
$\$20\,000 < x \leq \$25\,000$	127
$\$25\,000 < x \leq \$30\,000$	43
$\$30\,000 < x \leq \$35\,000$	7

    a) Draw a cumulative frequency graph for the salaries data

    b) Use your graph to estimate the median and quartiles.

    c) Estimate the proportion of the employees who earn less than half the median salary?

14. A computer can generate random numbers which are either 0 or 2. On a particular occasion, it generates a set of numbers which consists of 23 zeros and 17 twos. Find the mean and variance of this set of 40 numbers. [4]

Cambridge International AS and A Level Mathematics 9709, Paper 6 Q1 October/November 2003

15. The ages, x years, of 18 people attending an evening class are summarised by the following totals: $\sum x = 745$, $\sum x^2 = 33951$.

   i) Calculate the mean and standard deviation of the ages of this group of people. [3]

   ii) One person leaves the group and the mean age of the remaining 17 people is exactly 41 years.

Find the age of the person who left and the standard deviation of the ages of the remaining 17 people [4]

Cambridge International AS and A Level Mathematics 9709, Paper 6 Q4 October/November 2004

16. Two cricket teams kept records of the number of runs scored by their teams in 8 matches. The scores are shown in the following table.

Team A	150	220	77	30	298	118	160	57
Team B	166	142	170	93	111	130	148	86

   i) Find the mean and standard deviation of the scores for team A. [2]

The mean and standard deviation for team B are 130.75 and 29.63 respectively.

   ii) State with a reason which team has the more consistent scores. [2]

Cambridge International AS and A Level Mathematics 9709, Paper 6 Q1 May/June 2004

17. The values, x, in a particular set of data are summarised by $\sum (x - 25) = 133$, $\sum (x - 25)^2 = 3762$.

The mean, $\bar{x}$, is 28.325

   i) Find the standard deviation of x. [4]

   ii) Find $\sum x^2$ [2]

Cambridge International AS and A Level Mathematics 9709, Paper 61 Q2 October/November 2011

18. The lengths of time in minutes to swim a certain distance by the members of a class of twelve 9-year-olds and by the members of a class of eight 16-year-olds are shown below.

9-year-olds:	13.0	16.1	16.0	14.4	15.9	15.1	14.2	13.7	16.7	16.4	15.0	13.2
16-year-olds:	14.8	13.0	11.4	11.7	16.5	13.7	12.8	12.9				

   i) Draw a back-to-back stem-and-leaf diagram to represent the information above. [4]

   ii) A new pupil joined the 16-year-old class and swam the distance. The mean time for the class of nine pupils was now 13.6 minutes. Find the new pupil's time to swim the distance. [3]

Cambridge International AS and A Level Mathematics 9709, Paper 61 Q4 May/June 2007

19. In a survey, people were asked how long they took to travel to and from work, on average. The median time was 3 hours 36 minutes, the upper quartile was 4 hours 42 minutes and the interquartile range was 3 hours 48 minutes. The longest time taken was 5 hours 12 minutes and the shortest time was 30 minutes.

   i) Find the lower quartile. [2]

   ii) Represent the information by a box-and-whisker plot, using a scale of 2 cm to represent 60 minutes. [4]

Cambridge International AS and A Level Mathematics 9709, Paper 6 Q3 October/November 2006

20. The weights in grams of a number of stones, measured correct to the nearest gram, are represented in the following table.

Weight (grams)	1–10	11–20	21–25	26–30	31–50	51–70
Frequency	$2x$	$4x$	$3x$	$5x$	$4x$	x

A histogram is drawn with a scale of 1 cm to 1 unit on the vertical axis, which represents frequency density. The 1–10 rectangle has height 3 cm.

i) Calculate the value of x and the height of the 51–70 rectangle. [4]

ii) Calculate an estimate of the mean weight of the stones. [3]

Cambridge International AS and A Level Mathematics 9709, Paper 61 Q4 October/November 2010

21. There are 5000 schools in a vertain country. The cumulative frequency table shows the number of pupils in a school and the corresponding number of schools.

Number of pupils in a school	≤100	≤150	≤200	≤250	≤350	≤450	≤600
Cumulative frequency	200	800	1600	2100	4100	4700	5000

i) Draw a cumulative frequency graph with a scale of 2 cm to 100 pupils on the horizontal axis and a scale of 2 cm to 1000 schools on the vertical axis. Use your graph to estimate the median number of pupils in a school. [3]

ii) 80% of the schools have more than n pupils. Estimate the value of n correct to the nearest ten. [2]

iii) Find how many schools have between 201 and 250 (inclusive) pupils. [1]

iv) Calculate an estimate of the mean number of pupils per school. [4]

Cambridge International AS and A Level Mathematics 9709, Paper 61 Q6 May/June 2011

22. The lengths of the diagonals in metres of the 9 most popular flat screen TVs and the 9 most popular conventional TVs are shown below.

Flat screen:	0.85	0.94	0.91	0.96	1.04	0.89	1.07	0.92	0.76
Conventional:	0.69	0.65	0.85	0.77	0.74	0.67	0.71	0.86	0.75

i) Represent this informaiton on a back-to-back stem-and-leaf diagram. [4]

ii) Find the median and the interquartile range of the lengths of the diagonals of the 9 conventional TVs. [3]

iii) Find the mean and standard deviation of the lengths of the diagonals of the 9 flat screen TVs. [2]

Cambridge International AS and A Level Mathematics 9709, Paper 61 Q5 May/June 2012

23. Prices in dollars of 11 caravans in a showroom are as follows.

16800 18500 17700 14300 15500 15300
16100 16800 17300 15400 16400 [3]

i) Represent these prices by a stem-and-leaf diagram. [3]

ii) Write down the lower quartile of the prices of the caravans in the showroom. [1]

iii) 3 different caravans in the showroom are chosen at random and their prices are noted. Find the probability that 2 of these prices are more than the median and 1 is less than the lower quartile. [3]

Cambridge International AS and A Level Mathematics 9709, Paper 61 Q4 October/November 2012

24. As part of a data collection exercise, members of a certain school year group were asked how long they spent on their Mathematics homework during one particular week. The times are given to the nearest 0.1 hour. The results are displayed in the following table.

Time spent (*t* hours)	$0.1 \le t \le 0.5$	$0.6 \le t \le 1.0$	$1.1 \le t \le 2.0$	$2.1 \le t \le 3.0$	$3.1 \le t \le 4.5$
Frequency	11	15	18	30	21

i) Draw, on graph paper, a histogram to illustrate this informaiton. [5]

ii) Calculate an estimate of the mean time spent on their Mathematics homework by members of this year group. [3]

Cambridge International AS and A Level Mathematics 9709, Paper 6 Q5 May/June 2008

25. The pulse rates, in beats per minute, of a random sample of 15 small animals are shown in the following table

115	120	158	132	125
104	142	160	145	104
162	117	109	124	134

i) Draw a stem-and-leaf diagram to represent the data. [3]

ii) Find the median and the quartiles. [2]

iii) On graph paper, using a scale of 2 cm to represent 10 beats per minute, draw a box-and-whisker plot of the data. [3]

Cambridge International AS and A Level Mathematics 9709, Paper 6 Q5 October/November 2008

26. A study of the ages of car drivers in a certain country produced the results shown in the table.

Percentage of drivers in each age group			
	Young	Middled-aged	Elderly
Males	40	35	25
Females	20	70	10

Illustrate these results diagrammatically. [4]

Cambridge International AS and A Level Mathematics 9709, Paper 6 Q1 October/November 2005

27. The following table give the marks, out of 75, in a pure mathematics examination taken by 234 students.

Marks	1–20	21–30	31–40	41–50	51–60	61–75
Frequency	40	34	56	54	29	21

i) Draw a histogram on graph paper to represent the results. [5]

ii) Calculate estimates of the mean mark and the standard deviation. [4]

Cambridge International AS and A Level Mathematics 9709, Paper 62 Q6 October/November 2009

28. ii) The following data represent the daily ticket sales at a small theatre during three weeks.

52, 73, 34, 85, 62, 79, 89, 50, 45, 83, 84, 91, 85, 84, 87, 44, 86, 41, 35, 73, 86.

a) Construct a stem-and-leaf diagram to illustrate the data. [3]

b) Use your diagram to find the median of the data. [1]

Cambridge International AS and A Level Mathematics 9709, Paper 6 Q1 ii May/June 2003

Maths in real-life

Seeing the wood and the trees

The world is a complex place and almost all interesting problems in society involve multiple factors. More often than not relationships are not linear and there are interactions between effects.

Statistics is becoming more complex as technology enables:

- more data to be collected

- analysis of large quantities of data to be undertaken on a useful timescale – who wants an accurate weather forecast based on meteorological data which is six days old?

- new data analytical techniques which rely on the data properties and not on assumptions in models which are only approximations to the distribution.

- new ways to visualise complex relationships and to display more variables simultaneously.

This display shows vividly how the health and wealth of nations is strongly correlated, but it contains much more information as well. The colour identifies the continent and the size of the bubble shows the population of the country.

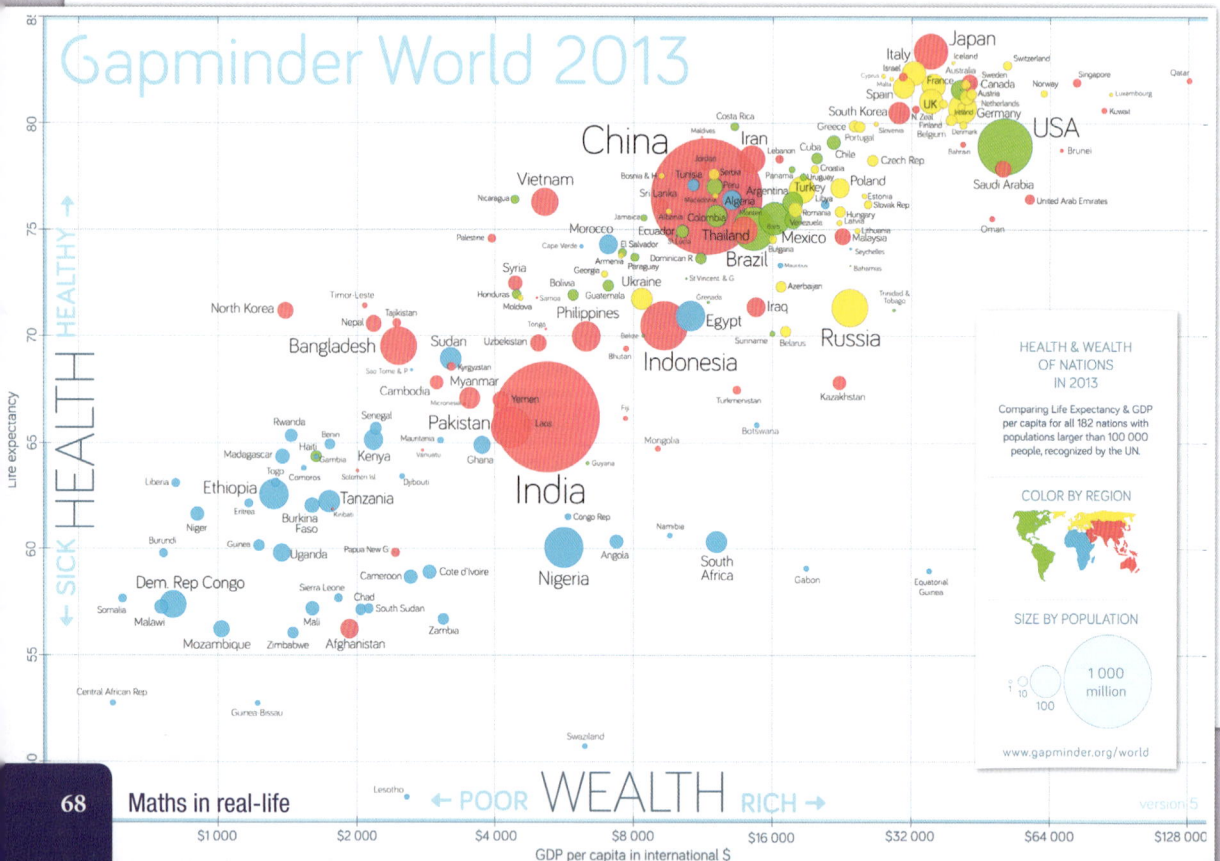

Gapminder World 2013

HEALTHY

HEALTH

SICK

Life expectancy

HEALTH & WEALTH
OF NATIONS
IN 2013

Comparing Life Expectancy & GDP per capita for all 182 nations with populations larger than 100 000 people, recognized by the UN.

COLOR BY REGION

SIZE BY POPULATION

1 000 million

www.gapminder.org/world

← POOR WEALTH RICH →

GDP per capita in international $

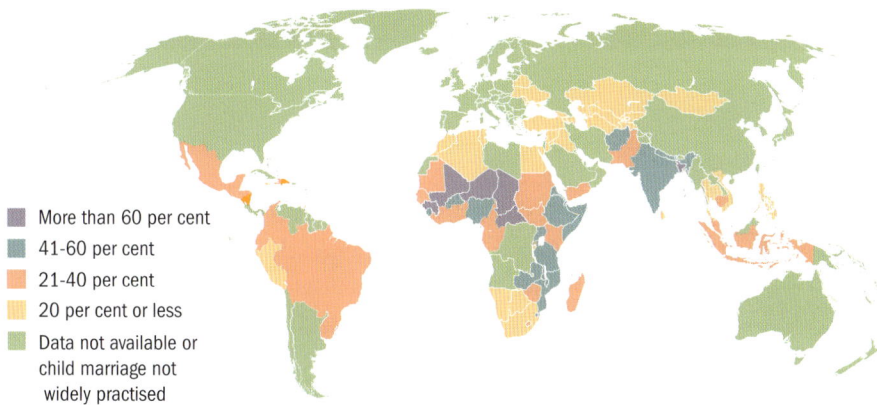

More than 60 per cent
41-60 per cent
21-40 per cent
20 per cent or less
Data not available or child marriage not widely practised

Geographical Information Systems (GiS) are being used increasingly to display a spatial component of data where this communicates an important feature of the full data set. These statistics produced by UNICEF show the proportion of women aged 20 – 24 in each country who were married or in union before the age of 18.

Infographics is a buzzword when talking about representing data. It means communicating information visually through use of graphics (not always graphs – but any sort of images). By definition, infographics should communicate the information quickly and clearly, for example the graphic on the right gives us some quick and interesting facts on languages.

These infographics are an easy way to get an overview of the information, although, as we saw in the introduction to Chapter 2, we should be careful not to rely too heavily on summary statistics as this may not give us the whole picture.

The development of easy-to-use, free tools have made the creation of infographics more readily available to the public. Through social media, such as Facebook and Twitter, Infographics can now be shared and spread quickly among many people, worldwide.

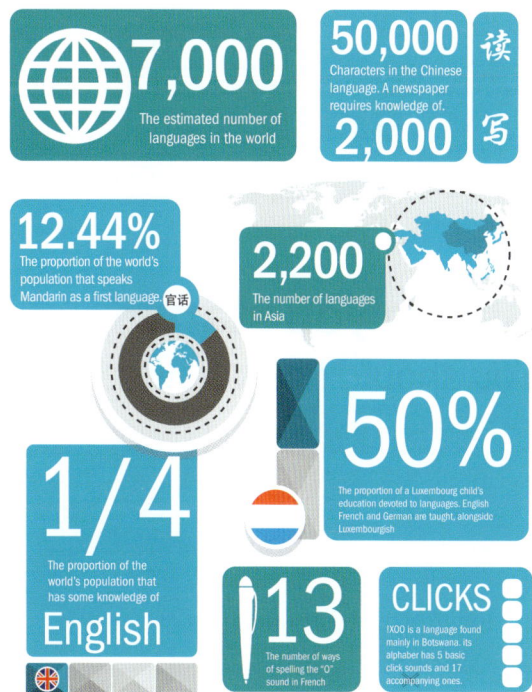

7,000
The estimated number of languages in the world

50,000
Characters in the Chinese language. A newspaper requires knowledge of.
2,000
读 写

12.44%
The proportion of the world's population that speaks Mandarin as a first language. 官话

2,200
The number of languages in Asia

1/4
The proportion of the world's population that has some knowledge of
English

50%
The proportion of a Luxembourg child's education devoted to languages. English French and German are taught, alongside Luxembourgish

13
The number of ways of spelling the "O" sound in French

CLICKS
!XOO is a language found mainly in Botswana. Its alphabet has 5 basic click sounds and 17 accompanying ones.

4 Probability

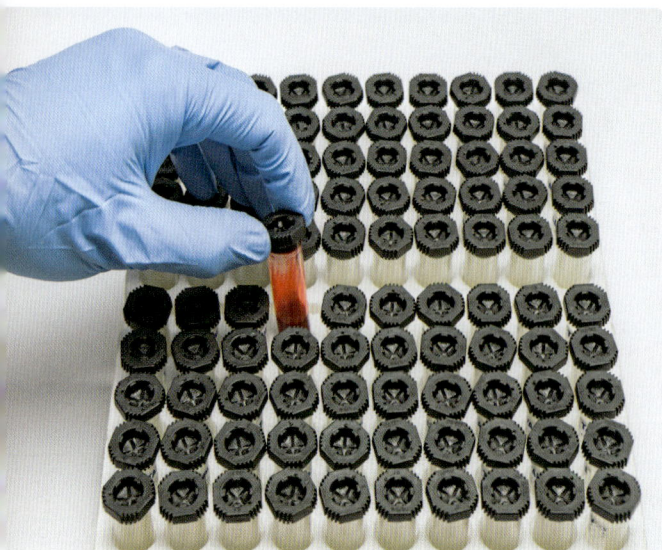

Test for diseases can return false positives (a positive result when the patient does not have the disease) as well as positive results for patients who do have the disease. Conditional probability allows doctors to have some idea of the likelihood the patient really has the disease when a positive results occurs. Misunderstanding of conditional probabilities has caused serious miscarriages of justice; for example, in the UK, Sally Clark was wrongly convicted of murdering her two sons who died from sudden death infant syndrome.

Objectives

- Evaluate probabilities in simple cases by means of enumeration of equiprobable elementary events (e.g. for the total score when two fair dice are thrown).
- Use addition and multiplication of probabilities, as appropriate, in simple cases.
- Understand the meaning of exclusive and independent events, and calculate and use conditional probabilities in simple cases, e.g. situations that can be represented by means of a tree diagram.

Before you start

You should know how to:

1. Work with fractions: e.g. Calculate

   a) $\frac{7}{12} + \frac{1}{3}$ b) $\frac{1}{6} \times \frac{1}{2}$

   a) $\frac{7}{12} \times \frac{1}{3} = \frac{7+4}{12} = \frac{11}{12}$ b) $\frac{1}{6} \times \frac{1}{2} = \frac{1}{12}$

2. Identify the basic outcomes of a simple experiment

   e.g. how many different pairs of letters can be made from the word DICE?

   DI, DC, DE, IC, IE, CE

Skills check:

1. Calculate

   a) $\frac{1}{4} \times \frac{2}{3}$ b) $\frac{1}{4} \times \frac{2}{3}$

2. List the possible outcomes when a coin is tossed twice

4.1 Basic concepts and language of probability

A **probability experiment** has outcomes which occur unpredictably.

Imagine an experiment where you roll a die 500 times, and you see 74 'fives'. The **relative frequency** or **experimental probability** of throwing a five in the experiment is $\frac{74}{500} = 0.148$.

However, provided the die is fair, the **theoretical probability** of throwing a five is $\frac{1}{6} = 0.166....$ So, if the die is fair, 'on average' you would see $\frac{1}{6}$ of 500 ≈ 83 fives.

> The probability is $\frac{1}{6}$ because there are six possible scores when rolling a die.

It might be argued that the die used in the experiment is slightly biased against rolling a five, but the evidence for this is not very strong. When the experiment is recreated, the number of fives varies considerably.

To be more confident, you should roll the die more times: 7416 fives in 50 000 rolls would be stronger evidence that the die was not fair than 74 fives in 500 rolls, even though the proportion of fives is almost the same.

> The relative frequency of an event happening can be used as an estimate of the probability of that event happening. The estimate is more likely to be close to the true probability if the experiment has been carried out a large number of times.

An **event** is a set of possible outcomes from an experiment.

So for rolling a die, you could say:

A is the event that a five is rolled.

B is the event that an even number is rolled.

C is the event that an odd number is rolled.

$A \cup B$ is the **union** of events A and B.
This means A or B or both can happen.
Rolling 2, 4, 5, 6 are the outcomes which satisfy $A \cup B$.

$A \cap C$ is the **intersection** of events A and C.
This means both A and C have to happen.
Rolling 5 is the only outcome that satisfies $A \cap C$.

A' means the event 'A does not happen'.
This is the **complementary event**, and $P(A') = 1 - P(A)$.

Rolling 1, 2, 3, 4, 6 are the outcomes which satisfy A'.

> Sometimes the complementary probability is much easier to work out directly

Note that $B \cap C$ has no outcomes satisfying it – there are no numbers which are both even and odd. This can be written as $B \cap C = \emptyset$ or $B \cap C = \{\}$ and is referred to as the **null set** or **empty set**.

Exercise 4.1

1. A normal die is thrown. The events A, B, C and D are defined as:

   A: A factor of 4 is seen. B: A square number is seen.

   C: A prime number is seen. D: A multiple of 3 is seen.

   a) For each event, A, B, C and D, write down the outcomes which satisfy it.

   b) Give the probability of each event A, B, C and D.

   c) List the outcomes which satisfy $A \cap C$.

   d) Write down $P(A \cap C)$.

   e) Find $P(A \cap D)$.

   f) Find $P(A \cup B)$.

   > $P(A \cap C)$ stands for 'the probability that A and C *both* happen.'

2. A letter is chosen at random from the word CAMBRIDGE. The events A, B, C and D are defined as:

   A: A vowel is chosen.

   B: The letter B is chosen.

   C: A letter in the first half of the alphabet is chosen.

   D: A letter is chosen which has only one letter beside it.

   a) Describe the event A' in words.

   b) For each event, A, B, C and D, write down the outcomes which satisfy it.

   c) Give the probability of each event A, B, C and D.

   d) List the outcomes which satisfy $A \cap C$.

   e) Write down $P(A \cap C)$.

   f) Find $P(A \cap D)$.

   g) Find $P(A \cup B)$.

3. a) Toss a coin 20 times and count the number of times it shows heads.

   b) Toss the coin another 20 times and count the number of times it shows heads.

   c) How many times would you 'expect' to see heads in 20 tosses?

   d) Did you get this in both sets of 20 coin tosses?

   e) If you have access to a number of other people's results as well – how often do you see the 'expected number' of heads?

4. a) Roll a die 30 times and count the number of times it shows a five.

   b) How many times would you 'expect' to see a five in 30 rolls?

   c) Roll the die another 20 times and count the number of times it shows a five.

   d) How many times would you 'expect' to see a five in 20 rolls?

   e) If you have access to a number of other people's results as well – how often do you see the 'expected' number of fives:

        i) in 30 throws?   ii) in 20 throws?

4.2 Two (or more) events

There are many contexts in which you are interested in the outcome of more than one thing at a time – that is, of a **compound event**. In simple cases it may be possible to make a list of all the possible outcomes. In order to make it easier to keep track, it is common to write the outcomes as ordered pairs (or larger groups, if more terms are needed) inside brackets, ( ), and to put the whole list inside curly brackets, { }. This list is called the **sample space** of the experiment.

So for tossing a coin twice the sample space will be: $\{(H, H); (H, T); (T, H); (T, T)\}$

In a context where two things happen and you combine them according to some rule, it is often helpful to construct a table showing the possible outcomes. This is sometimes called a **possibility space diagram** or a **sample space diagram**.

For example, a situation in which two dice are thrown and the sum of the scores is taken is represented in the **two-way table** shown here.

Sum	1	2	3	4	5	6
1	2	3	4	5	6	7
2	3	4	5	6	7	8
3	4	5	6	7	8	9
4	5	6	7	8	9	10
5	6	7	8	9	10	11
6	7	8	9	10	11	12

It helps to write a sample space list in a logical, methodical order. For tossing a coin three times the sample space can be written as: $\{(H, H, H); (H, H, T); (H, T, H); (H, T, T); (T, H, H); (T, H, T); (T, T, H); (T, T, T)\}$.

Within this sample space, the list for tossing a coin twice appears two times – first in red with 'H' in front and then in blue with 'T' in front.

This list of eight outcomes for tossing a coin three times can then form the basis of the sixteen outcomes for tossing a coin four times. This is created by putting an extra 'H' in front and then an extra 'T' in front, which in turn can be used for the list for tossing a coin five times, and so on.

The blue cell represents the ordered pair $(1, 5)$ – as indicated by the row and column headings – but you can then insert the **value** of interest inside the cell; in this case you can insert the sum of the two scores, which is 6.

You can use the table to work out the probability of getting a sum of 5 (the instances of a sum of 5 are in grey cells):

$$P(\text{sum} = 5) = \frac{4}{36}$$

The advantage of this approach is that the (row, column) pair identifies what is actually seen in the experiment, while the contents of the cell show the result of this ordered pair of outcomes, according to the particular rule to be applied.

This rule can change. For example, two dice are again thrown, but this time the higher of the scores on the two dice is taken:

High	1	2	3	4	5	6
1	1	2	3	4	5	6
2	2	2	3	4	5	6
3	3	3	3	4	5	6
4	4	4	4	4	5	6
5	5	5	5	5	5	6
6	6	6	6	6	6	6

You can use this table to work out the probability of the higher score being 5:

$$P(\text{high} = 5) = \frac{9}{36}$$

> A two-way table can be used to show the possible outcomes of a compound event such as throwing two dice.

Exercise 4.2

1. There are three starters on a restaurant menu: vegetable pakora (V), onion bhaji (O) and chicken tikka (C). Jen and Kay each order a starter.

   List the sample space of possible orders.

2. In question **1**, if there is only one of each of the starters left, list the sample space of possible orders.

3. A set of six cards show the numbers 1 to 6. Two cards are taken at random.

   Copy and complete the sample space diagram to show the sum of the numbers on the cards.

Sum	1	2	3	4	5	6
1			4			
2						
3						
4						
5						
6		8				

   Find the probability that the total score is

   a) 5   **b)** 4   **c)** 2.

4. A coin is tossed and a die is thrown.

   a) List all the possible outcomes in the sample space.

   A head scores 1 and a tail scores 2.

   b) Construct a sample space diagram to show the total scores for this experiment.

5. Two dice are thrown.

   a) Construct a sample space diagram to show the product of the scores on the two dice.

   b) Find the probability that the score is:

   i) 3   **ii)** 5   **iii)** 6   **iv)** 10.

6. Two dice are thrown.

   a) Construct a sample space diagram to show the lower of the scores on the two dice.

   b) Find the probability that the score is

   i) 3   **ii)** 5   **iii)** 6.

7. Two dice are thrown.

   a) Construct a sample space diagram to show the (unsigned) difference between the scores on the two dice.

   b) Find the probability that the score is

   i) 3   **ii)** 5   **iii)** 6.

4.3 Tree diagrams

Consider a bag which has three red and five blue beads in it.

If you take a bead out at random, and then take out another without replacing the first, you can represent the possible outcomes in a **possibility tree diagram**, as shown on the right.

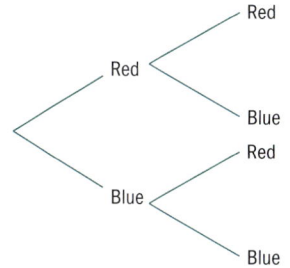

You can put probabilities on the branches to complete the diagram:

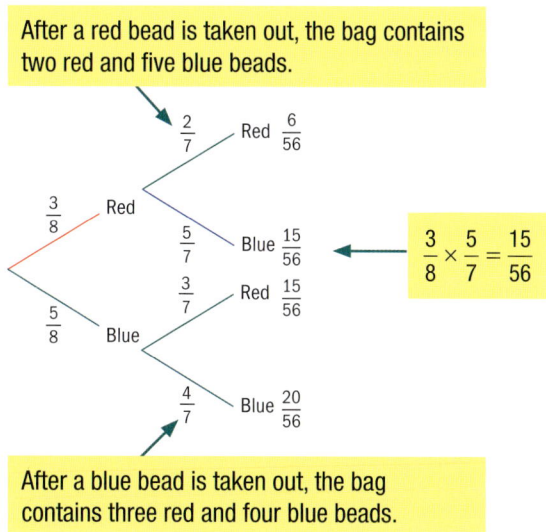

After a red bead is taken out, the bag contains two red and five blue beads.

$$\frac{3}{8} \times \frac{5}{7} = \frac{15}{56}$$

After a blue bead is taken out, the bag contains three red and four blue beads.

P(both beads the same colour) = P(both beads red)
+ P(both beads blue)

$$= \left(\frac{3}{8} \times \frac{2}{7}\right) + \left(\frac{5}{8} \times \frac{4}{7}\right)$$

$$= \frac{6}{56} + \frac{20}{56}$$

$$= \frac{26}{56}$$

$$= \frac{13}{28}$$

Think × for 'and', and + for 'or'.

So	red	*and*	red	*or*	blue	*and*	blue
	= red	×	red	+	blue	×	blue

Sampling with replacement

If the first bead was returned to the bag before the second
one was selected, the probabilities of red or blue would be
$\frac{3}{8}$ and $\frac{5}{8}$ at the second stage, as well as the first:

Now,

P(both beads the same colour)
= P(both beads red) + P(both beads blue)

$$= \frac{9}{64} + \frac{25}{64}$$

$$= \frac{34}{64}$$

If the first and second bead are returned to the bag, and a third
bead is taken out, the tree diagram will look like this:

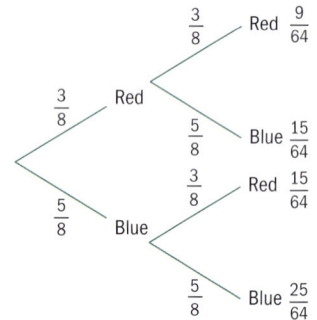

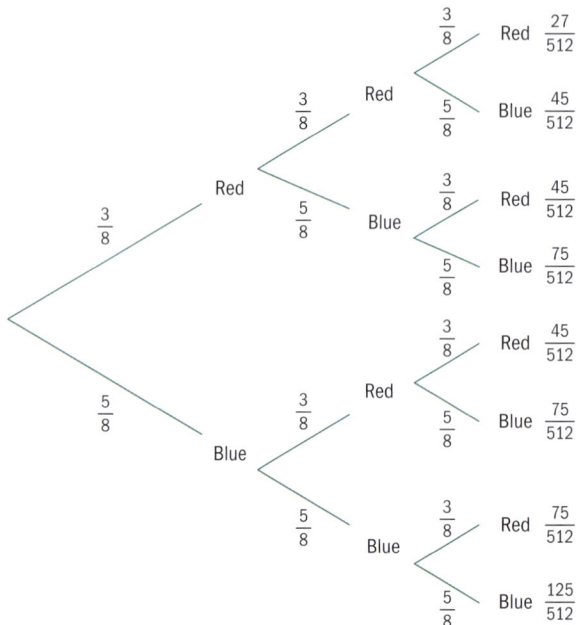

Example 1

A disease is known to affect 1 in 10 000 people. It can be fatal, but it is treatable if it is detected early.

A screening test for the disease shows a positive result for 99% of people with the disease.

The test shows positive for 2% of people who do not have the disease.

For a population of one million people

a) how many would you expect to have the disease and test positive

b) how many would you expect to test positive?

> A screening test will detect the presence in people's bodies of substances which are almost always present with the disease, but which also occur naturally in a small proportion of people.

First draw a tree diagram:

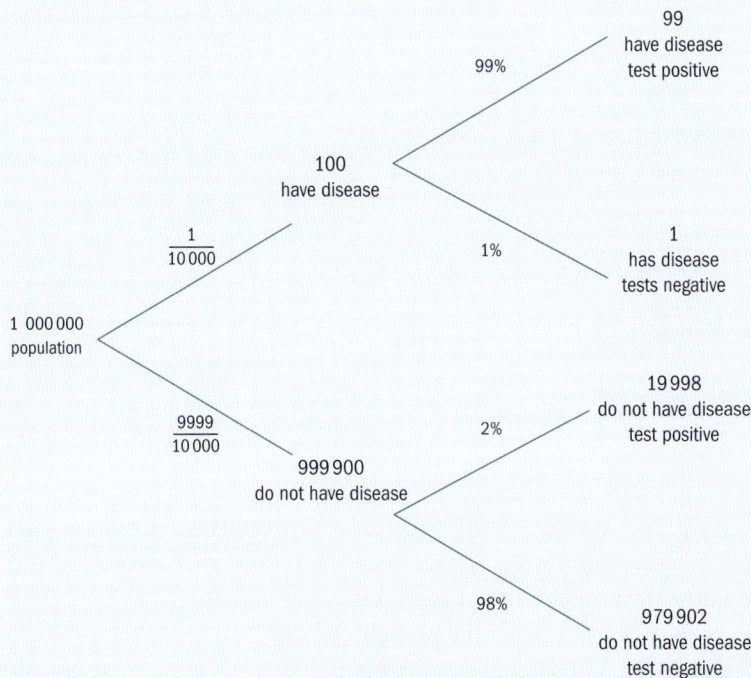

So there are likely to be about 99 positive tests from people with the disease, and about 99 + 19 998 = 20 097 positive tests altogether. In this scenario, fewer than 1 in 200 positive tests is from a person with the disease, despite the test being very accurate.

More complex tree diagrams

Example 2

There are four red, three green and five blue discs in a bag.

Find the probability that two discs the same colour are drawn.

....

First draw a tree diagram – there are three outcomes at each stage:

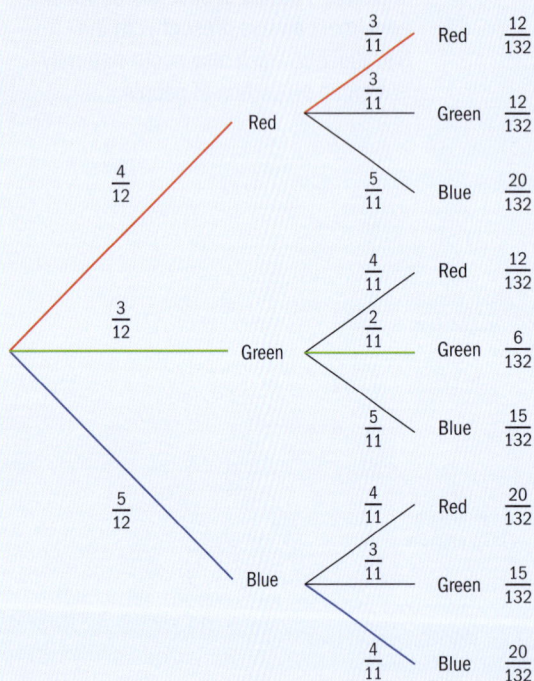

> Even though some fractions may be simplified, it is much easier to keep the same denominator for the probabilities at each stage

$$P(\text{same colour}) = \frac{12}{132} + \frac{6}{132} + \frac{20}{132} = \frac{38}{132} = \frac{19}{66}$$

> Tree diagrams are useful when you know the probabilities of each stage of compound events. You multiply along the branches to get the probability of a pathway, and the probabilities of different pathways can be added.

Exercise 4.3

1. A bag contains five blue and three green balls.

   A ball is chosen at random and the colour noted, then the ball is returned to the bag.

   A second ball is chosen.

   a) Find the probability that the two balls are different colours.

   b) If the first ball is not returned to the bag, what is the probability the two balls are different colours?

2. At a gym, 60% of the members are men. One third of the men use the gym at least once a week. Three-quarters of the women use the gym at least once a week.

A member is chosen at random. Find the probability that

a) it is a man who does not use the gym at least once a week

b) it is a person who uses the gym at least once a week.

3. For a person living in a particular town, the probability that during a period of one year they will start a new job is 0.06. The probability that they will be fired from a job is 0.03.

Assuming these events are independent, draw a tree diagram to represent this information.

Find the probability that during one year a randomly selected person living in the town has

a) neither of these events happen

b) exactly one of these events happen

c) both of these events happen.

4. A coin is tossed three times. Find the probability that

a) it shows heads on all three tosses

b) it shows the same on all three tosses

c) it does not show the same on two successive tosses.

5. A bag contains ten counters: four white, three green and three red. Counters are removed one at a time at random, without replacement. Find the probability that

a) the first counter drawn is red

b) the first three counters drawn are all white

c) the first three counters drawn are all different colours.

4.4 Conditional probability

In Example 1 you considered a screening test for a disease and found that, even though the test is very accurate, the fact that the disease is rare – there are many more people who do not suffer from it than people who do – means that fewer than 1 in 200 of those testing positive actually have the disease.

> In situations like this it is important that a person testing positive is not told that they have the disease based only on the test result

19 998 out of 20 097 positive test results were from healthy people, so the **conditional probability** that somebody is healthy **given that** they have a positive result is

$$P(\text{Healthy} \mid \text{positive test result}) = \frac{19\,998}{20\,097} \approx 0.9951$$

The conditional probability of an event A occurring given that an event B has already occurred can be written as $P(A|B)$.

Example 3

A medical centre encourages elderly people to have a flu vaccination each year. The vaccination reduces the likelihood of getting flu from 40% to 10%.

If 45% of the elderly people visiting the centre have the vaccination, find the probability that an elderly person chosen at random

a) gets flu   **b)** had the vaccination, given that they get flu.

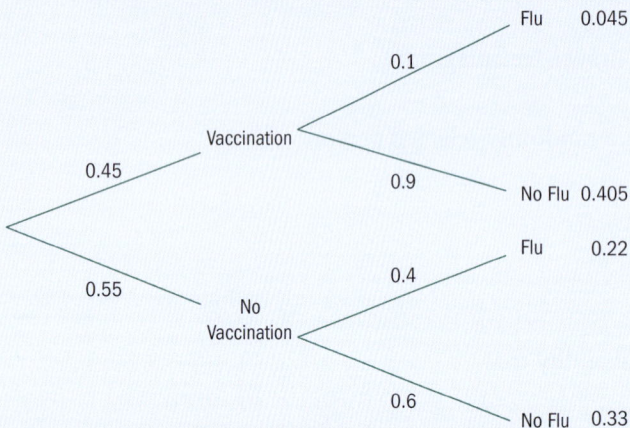

Flu   0.045

0.1

Vaccination

0.45

0.9

No Flu   0.405

Flu   0.22

0.4

0.55   No Vaccination

0.6

No Flu   0.33

> The symbol | stands for 'given that', i.e. $P(V\,|\,F)$ stands for the probability of V, given that F has already happened.

a) $P(F) = 0.045 + 0.22 = 0.265$ **b)** $P(V\,|\,F) = \dfrac{P(V \cap F)}{P(F)} = \dfrac{0.045}{0.265} = 0.170$ (to 3 s.f.)

The conditional probability of V given F is $P(V|F) = \dfrac{P(V \cap F)}{P(F)}$.

> Be careful to work out the probability of both V and F happening directly and not from the probabilities of V and F happening individually.

Exercise 4.4

1. From a sample taken, 95% of car drivers wear seat belts, 60% of car drivers involved in serious accidents die if they are not wearing a seat belt, and 80% of those that do wear a seat belt survive.

   a) Draw a tree diagram to show this information.

   b) What is the probability that a driver in a serious accident did not wear a seat belt and survived?

2. A shopkeeper buys one third of his stock of light bulbs from Company X, and the rest from Company Y.

   An independent report states that 3% of light bulbs from Company X are faulty and that 2% from Company Y are faulty.

   a) If the shopkeeper chooses a bulb at random from his stock and tests it, what is the probability that it is faulty?

   b) If the bulb is faulty, what is the probability that it came from Company Y?

3. The homework diaries and completed homework of two students, A and B, are examined.

   There is a probability of 0.4 that Student A does not make a note in her diary of the homework set. She always does the homework if it is written in her diary, but never does the homework if it is not written in.

   There is a probability of 0.8 that Student B writes the homework given in her diary. When she does this, she will do the homework 90% of the time. If she has nothing written in her diary then she checks with a friend, who knows what the homework is 50% of the time; Student B always does the homework if she is told what it is by her friend.

   Draw a tree diagram representing this information.

   a) Find the probability that Student B does her homework on a particular night.

   b) Find the probability that both students do their homework on a particular night.

   c) If one piece of homework was given to a student but wasn't done, find the probability that it was given to Student A.

4. In a school there are 542 students, of whom 282 are girls. Of these students, 364 walk to school, of whom 153 are girls. Find the probability that a student chosen at random

   a) is a boy

   b) is a boy who does not walk to school

   c) does not walk to school, given that they are a boy

   d) is a girl, given that they walk to school.

5. There are 173 students in one year in a school. In the school, 25 students play hockey, and of these 7 are in the school's hockey team.

   Find the probability that a student chosen from the year at random

   a) plays hockey

   b) plays in the school's team, given that they play hockey.

6. Of the employees in a large factory one sixth travel to work by bus, one third by train, and the rest by car. Those travelling by bus have a probability of $\frac{1}{4}$ of being late, those by train will be late with probability $\frac{1}{5}$, and those by car will be late with probability $\frac{1}{10}$

   Draw and complete a tree diagram to show this information. Calculate the probability that an employee chosen at random will be late.

7. An insurance company separates car drivers into three categories: Category X is 'low risk', and this category represents 20% of drivers who are insured with the company; Y is 'moderate risk' and represents 70% of drivers insured; Z is 'high risk'.

The probability that a Category X driver has one or more accidents in a 12-month period is 2%, and the corresponding probabilities for drivers in Categories Y and Z are 5% and 9%, respectively.

a) Find the probability that a driver insured with the company, chosen at random, is assessed as a Category Y risk and has one or more accidents in a 12-month period.

b) Find the probability that a driver insured with the company, chosen at random, has one or more accidents in a 12-month period.

c) If a customer has an accident in a 12-month period, what is the probability that they are a Category Y driver?

8. Two identical bags each contain 12 discs, which are identical except for colour. Bag A contains 6 red and 6 blue discs. Bag B contains 8 red and 4 blue discs.

a) A bag is selected at random and a disc is selected from it. Draw a tree diagram illustrating this situation and calculate the probability that the disc drawn will be red.

b) The disc selected is returned to the same bag, along with another two the same colour, and another disc is chosen from that bag. Find the probability that

   i) it is the same colour as the first disc drawn

   ii) bag A was used, given that two discs the same colour have been chosen.

4.5 Relationships between events

Events are often connected to each other by some type of relationship. The following are the main types.

> **Note:** ⇔ means 'implies and is implied by', i.e. $p \Leftrightarrow q$ means if p, then q; and if q, then p (p is equivalent to q).

Independence

Two events, V and F, are independent if the outcome of V does not affect the outcome of F, and vice versa.
$$P(V \mid F) = P(V) \Leftrightarrow V \text{ and } F \text{ are independent and,}$$
$$P(F \mid V) = P(F) \Leftrightarrow F \text{ and } V \text{ are independent.}$$

> The probability of V occurring given that F has already occurred will just be the probability of V

Example 4

A set of 40 cards shows a number, from 1 to 10, of one of four geometrical symbols.

Circles and squares are shown in grey, rectangles and ellipses are blue.

The pack of cards is shuffled and the top card is turned over. Let C be the event 'the card shows a circle', F be the event 'the card shows 5 symbols' and G be the event 'the card is grey'.

a) Show that C and F are independent events.

b) Show that C and G are not independent events.

a) $P(C) = \dfrac{10}{40}$ (there are 10 circle cards)

$P(F) = \dfrac{4}{10}$ (there are 4 cards showing 5 symbols)

$P(C \cap F) = \dfrac{1}{40}$ (there is only one card with 5 circles)

$P(C \mid F) = \dfrac{P(C \cap F)}{P(F)} = \dfrac{\frac{1}{40}}{\frac{4}{40}} = \dfrac{1}{4} = P(C)$

Since $P(C \mid F) = P(C)$, C and F are independent events.

b) $P(G \mid C) = 1$, since if you know the card has circles then you know it is grey

$P(G) = 0.5$

So $P(G \mid C) \neq P(G)$, and the events G and C are not independent.

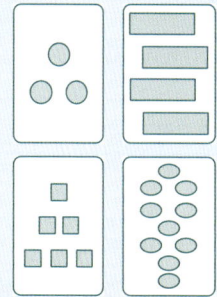

> Knowing that F has happened has given no additional information concerning whether C is likely to happen or not.

Addition law for probabilities

If we take all the outcomes that satisfy A, and then all the outcomes that satisfy B, then any outcomes which satisfy both will be double counted, so

$$P(A \cup B) = P(A) + P(B) - P(A \cap B)$$

That is, the probability of A or B can be found by adding the probabilities of A and B and then subtracting what has been double counted.

If A is the event a fair die shows a factor of 6 and B is the event that the fair die shows a square number, then 1, 2, 3 and 6 satisfy A, while 1 and 4 satisfy B. A and B is satisfied by 1, and A or B is satisfied by 1, 2, 3, 4 and 6.

> You can check this works in simple examples like events based on throwing a die, but logic says it is always going to be true, even in situations where events are not equally likely or where you cannot list all possible outcomes.

The addition law for probabilities is:
$$P(A \cup B) = P(A) + P(B) - P(A \cap B)$$

Mutually exclusive

Two events A and B are mutually exclusive if they cannot occur at the same time.
$$P(A \cup B) = P(A) + P(B)$$

But remember that the general relationship is
$P(A \cup B) = P(A) + P(B) - P(A \cap B)$

Exhaustive

A set of events is exhaustive if they cover all possible outcomes.

Example 5

Two fair dice are thrown. From the events listed, give two which, when taken together, are

a) mutually exclusive

b) exhaustive

c) not independent.

A: The two dice show the same number.

B: The sum of the two scores is at least 5.

C: At least one of the two numbers is a 5 or a 6.

D: The sum of the two scores is odd.

E: The largest number shown is a 6.

F: The sum of the two scores is less than 8.

a) A and D are mutually exclusive: if the two dice show the same number the sum has to be an even number.

b) B and F are exhaustive: the only outcomes not in B are that the sum is 2, 3 or 4 and these are all in F.

c) A and D are not independent (because they are mutually exclusive).
C and E are not independent, since $P(C|E) = 1$ (if E happens, then you know C must happen).

There are other possibilities for part (c).

Here are some other terms that you need to be aware of:

Partition is a group of sets which are exhaustive and mutually exclusive form a 'partition'. The whole outcome space has been split into disjoint events, so their probabilities total 1, and there is no overlap between any pair. Compound events can be evaluated simply by going through the group and seeing whether each set is to be included.

Complementary events are a special case of a partition; they are a two-event partition. If A and B are complementary then $P(B) = 1 - P(A)$. The simplest way of representing a complementary pair is as 'A and "not A"'.

Example 6

Over the course of a season, a hockey team play 40 matches, in different conditions, with the following results.

This is a **two-way table**

		Weather		Total
		Good	Poor	
Result	Win	13	6	19
	Draw	5	3	8
	Loss	7	6	13
Total		25	15	40

For a match chosen at random from the season:

G is the event 'Good weather'

W is the event 'Team wins'

D is the event 'Team draws'

L is the event 'Team loses'.

a) Find the probability in each case.

    i) $P(G)$ **ii)** $P(G \cap D)$ **iii)** $P(D|G)$

b) Are the events D and G independent?

....

a) i) During the season, 25 matches are played in good weather, so $P(G) = \dfrac{25}{40} = \dfrac{5}{8}$

    ii) $G \cap D$ is a draw played in good weather, and there are 5 of those, so $P(G \cap D) = \dfrac{5}{40} = \dfrac{1}{8}$

    iii) $P(D|G) = \dfrac{P(G \cap D)}{P(G)} = \dfrac{\frac{1}{8}}{\frac{5}{8}} = \dfrac{1}{5}$

However, note that W & G are not independent and L & G are not independent either.

b) Since $P(D|G) = P(D)$ the events D and G are independent.

Exercise 4.5

1. A and B are independent events. $P(A) = 0.7$, $P(B) = 0.4$

   Find:

   a) $P(A \cap B)$ b) $P(A \cup B)$ c) $P(A' \cap B)$

2. $P(A) = 0.7$, $P(B) = 0.4$, $P(A \cup B) = 0.82$.

   Show that A and B are independent.

3. $P(A) = 0.5$, $P(B|A) = 0.6$, $P(B') = 0.7$

   Show that A' and B are mutually exclusive.

4. X and Y are independent events with $P(X) = 0.4$ and $P(Y) = 0.5$.

   a) Write down $P(X|Y)$.   b) Write down $P(Y|X)$.   c) Calculate $P(X' \cap Y)$.

5. The results of a survey of colours and types of cars are shown in the table.

	Saloon	Hatchback
Silver	65	59
Black	27	22
Other	16	19

   One car is selected from the group at random.

   a) Find the probability that the selected car is

       i)   a silver hatchback

       ii)   a hatchback

       iii)   a hatchback, given that it is silver.

   b) Show that the type of car is not independent of its colour.

6. Consider the following possible outcomes of rolling a blue die and a white die:

   A: The total is 2.

   B: The white die shows a multiple of 2.

   C: The total is less than 10.

   D: The white die shows a multiple of 3.

   E: The total is greater than 7.

   F: The total is greater than 9.

   Which of the following pairs of events are exhaustive? Which are mutually exclusive?

   a) A, B b) A, D

   c) C, E d) C, F

   e) B, D f) A, E

7. The two sides of a coin are known as 'head' and 'tail'. Four unbiased coins are tossed together. Possible events are:

A: No heads.
B: At least one head.
C: No tails.
D: At least two tails.

Say whether each statement is true or false, giving a reason for your answer.

Recall that X' means 'not X'.

a) A and B are mutually exclusive.

b) A and B are exhaustive.

c) B and D are exhaustive.

d) A' and C' are mutually exclusive.

Summary exercise 4

EXAM-STYLE QUESTION

1. A bag contains eight purple balls and two pink balls. A ball is selected at random from the bag and its colour is recorded. The ball is not replaced. A second ball is selected at random and its colour is recorded.

a) Draw a tree diagram to represent this information.

Find the probability that

b) the second ball selected is purple

c) both balls selected are purple, given that the second ball selected is purple.

2. Walker's disease is a rare tropical disease, known to be present in only 0.1% of the population. A new screening test has been analysed, and there is a 98% probability of testing positive when the person tested has the disease, and only a 0.2% probability of testing positive when the person does not have the disease.

A person is selected at random from the population and given the screening test.

a) What is the probability that the person will test positive?

b) What is the probability that the person does not have the disease, given that they test positive?

c) Jane is a doctor who is unhappy with guidelines which say that patients should be told immediately if the test shows positive. Explain how she could use the answer to part (b) to argue that these guidelines are not appropriate.

EXAM-STYLE QUESTION

3. Two events A and B are mutually exclusive. $P(A) = \frac{1}{2}$, $P(B) = \frac{1}{3}$

a) Find $P(A \mid B)$.

b) Find $P(A \cup B)$.

c) Are events A and B independent? Provide a reason for your answer.

Probability 87

4. The events A and B are such that

    $P(A) = \frac{5}{12}$, $P(B) = \frac{2}{3}$ and $P(A' \cap B') = \frac{1}{12}$.

    a) Find:

        i) $P(A \cap B')$ ii) $P(A \mid B)$

        iii) $P(B \mid A)$

    b) State, giving a reason, whether or not A and B are

        i) mutually exclusive

        ii) independent.

5. A computer-based testing system gives the user a hard question if they got the previous question correct and an easy question if they got the previous question wrong. The first question is randomly chosen to be hard or easy.

    The probability of Benni getting an easy question right is $\frac{2}{3}$ and the probability he gets a hard question right is $\frac{1}{4}$.

    a) Draw a tree diagram to represent what can happen for the first two questions Benni gets in a test.

    b) Find the probability Benni gets his first two questions correct.

    c) Find the probability that the first question was hard, given that Benni got both of his first two questions correct.

6. In a factory, Machines X, Y and Z are all producing identical metal rods. Machine X produces 25% of the rods, Machine Y produces 45% and the rest are produced by Machine Z. The production of rods from Machines X, Y and Z are 4%, 5% and 2% defective, respectively.

    a) Draw a tree diagram to represent this information.

    b) Find the probability that a randomly selected rod is

        i) produced by Machine Y and not defective

        ii) not defective.

    c) Given that a randomly selected rod is not defective, find the probability that it was produced by Machine Y.

7. A golfer enters two tournaments. He estimates the probability that he wins the first tournament is 0.6, that he wins the second tournament is 0.4 and that he wins both tournaments is 0.35.

    a) Find the probability that he does not win either tournament.

    b) Show, by calculation, that winning the first tournament and winning the second tournament are not independent events.

    c) The tournaments are played in successive weeks. Explain why it would be surprising if these were independent events.

8. The events A and B are independent such that $P(A) = \frac{1}{2}$ and $P(B) = \frac{1}{3}$.
    Find:

    a) $P(A \cap B)$

    b) $P(A' \cap B')$

    c) $P(A \mid B)$.

9. A fair die has six faces, numbered 4, 4, 4, 5, 6, and 6. The die is rolled twice and the number showing on the uppermost face is recorded each time.

    Find the probability that the sum of the two numbers recorded is at least 14.

10. Events A, B and C are defined in the sample space S. The events A and B are independent.

Given that P(A) = 0.3, = 0.4 and P($A \cup B$) = 0.65, find

a) P(B)

b) P($A \mid B$),

A and C are mutually exclusive and P(C) = 0.5.

c) Find P($A \cup C$).

11. a) If A and B are two events which are statistically independent, write down expressions for P($A \cap B$) and P($A \cup B$) in terms of P(A) and P(B).

b) Each Friday, Anji and Katrina decide independently of one another whether to go to the cinema. On any given Friday, the probability of them both going to the cinema is $\frac{1}{3}$, and the probability that at least one of them goes is $\frac{5}{6}$.

Find the possible values for the probability that Anji goes to the cinema on a particular Friday.

12. Of the students who took English in a certain school one year, 60% of them took History, 30% of them took Biology, and 10% took both History and Biology.

One of the students taking English is chosen at random.

a) Find the probability that the student took neither History nor Biology.

b) Given that the student took exactly one of History and Biology, find the probability it was History.

13. Jamie is equally likely to attend or not to attend a training session before a football match. If he attends, he is certain to be chosen for the team which plays in the match. If he does not attend, there is a probability of 0.6 that he is chosen for the team.

i) Find the probability that Jamie is chosen for the team.

ii) Find the conditional probability that Jamie attended the training session, given that he was chosen for the team.

Cambridge International AS and A Level Mathematics 9709, Paper 6 Q2 May/June 2007

14. Data about employment for males and females in a small rural area are shown in the table.

	Unemployed	Employed
Male	206	412
Female	358	305

A person from this area is chosen at random. Let M be the event that the person is male and let E be the event that the person is employed.

i) Find P(M).

ii) Find P(M and E).

iii) Are M and E independent events? Justify your answer.

iv) Given that the person chosen is unemployed, find the probability that the person is female.

Cambridge International AS and A Level Mathematics 9709, Paper 6 Q5 May/June 2005

15. In a television quiz show Peter answers questions one after another, stopping as soon as a question is answered wrongly.

- The probability that Peter gives the correct answer himself to any question is 0.7.
- The probability that Peter gives a wrong answer himself to any question is 0.1.
- The probability that Peter decides to ask for help for any question is 0.2.

On the first occasion that Peter decides to ask for help he asks the audience. The probability that the audience gives the correct answer to any question is 0.95. This information is shown in the tree diagram below.

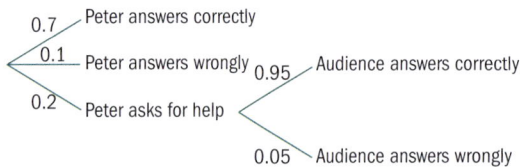

i) Show that the probability that the first question is answered correctly is 0.89.

On the second occasion that Peter decides to ask for help he phones a friend. The probability that his friend gives the correct answer to any question is 0.65.

ii) Find the probability that the first two questions are both answered correctly.

iii) Given that the first two questions were both answered correctly, find the probability that Peter asked the audience

Cambridge International AS and A Level Mathematics 9709, Paper 61 Q7 May/June 2010

16. Three friends, Rick, Brenda and Ali, go to a football match but forget to say which entrance to the ground they will meet at. There are four entrances, *A*, *B*, *C* and *D*. Each friend chooses an entrance independently.

- The probability that Rick chooses entrance *A* is $\frac{1}{3}$. The probabilities that he chooses entrances *B*, *C* or *D* are all equal.
- Brenda is equally likely to choose any of the four entrances.
- The probability that Ali chooses entrance *C* is $\frac{2}{7}$ and the probability that he chooses entrance *D* is $\frac{3}{5}$. The probabilities that he chooses the other two entrances are equal.

i) Find the probability that at least 2 friends will choose entrance *B*. [4]

ii) Find the Probability that the three friends will all choose the same entrance. [4]

Cambridge International AS and A Level Mathematics 9709, Paper 61 Q5 October/November 2010

17. When Ted is looking for his pen, the probability that it is in his pencil case is 0.7. If his pen is in his pencil case he always finds it. If his pen is somewhere else, the probability that he finds it is 0.2. Given that Ted finds his pen when he is looking for it, find the probability that it was in his pencil case. [4]

Cambridge International AS and A Level Mathematics 9709, Paper 61 Q2 May/June 2011

18. Bag A contains 4 balls numbered 2, 4, 5, 8.
Bag B contains 5 balls numbered 1, 3, 6, 8, 8.
Bag C contains 7 balls numbered. 2, 7, 8, 8, 8, 9.
One ball is selected at random from each bag.

  i) Find the probability that exactly two of the selected balls have the same number. [5]

 ii) Given that exactly two of the selected balls have the same number, find the probability that they are both numbered 2. [2]

iii) Event X is 'exactly two of the selected balls have the same number'. Event Y is 'the ball selected from bag A has number 2'. Showing your working. Determine whether events X and Y are independent or not. [2]

Cambridge International AS and A Level Mathematics 9709, Paper 61 Q7 October/November 2011

Chapter summary

- The relative frequency of an event happening can be used as an estimate of the probability of that event happening. The estimate is more likely to be close to the true probability if the experiment has been carried out a large number of times.

- A two-way table can be used to show the possible outcomes of a compound event such as throwing two dice.

- Tree diagrams are useful when you know the probabilities of each stage of compound events. You multiply along the branches to get the probability of a pathway, and the probabilities of different pathways can be added.

- The conditional probability of V given F is $P(V \mid F) = \dfrac{P(V \cap F)}{P(F)}$.

  Be careful to work out the probability of both V and F happening directly and not from the probabilities of V and F happening individually.

- Two events, V and F, are independent if $P(V \mid F) = P(V)$. That is, knowing that F has happened has given no information about the likelihood of V happening, and vice versa.

- The addition law for probabilities is:
  $$P(A \cup B) = P(A) + P(B) - P(A \cap B)$$

- Two events A and B are mutually exclusive if they cannot occur at the same time.
  $$P(A \cup B) = P(A) + P(B)$$

- A set of events is exhaustive if they cover all possible outcomes.

Probability distributions and discrete random variables

Simulations are now in widespread use to model what might be expected to happen if different strategies are used in dealing with a problem. For example, underground Metro and Subway stations around the world often have to undertake important engineering works. The use of random variables (both discrete and continuous) and their probability distributions is integral to developing good models of people's behavior so the maintenance work causes as little disruption and inconvenience as possible.

Objectives

- Be able to define a discrete random variable.
- Construct a probability distribution table relating to a given situation involving a discrete random variable X.
- Calculate $E(X)$, the mean or expected value of X.
- Calculate $Var(X)$, the variance of X.

Before you start

You should know how to:

1. Substitute into simple expressions, e.g. find the value of the expression
$$\frac{1}{x} + \frac{1}{x+1} + \frac{1}{x+2} \text{ when } x = 2$$
$$\frac{1}{2} + \frac{1}{(2+1)} + \frac{1}{(2+2)} = \frac{1}{2} + \frac{1}{3} + \frac{1}{4}$$
$$= \frac{6+4+3}{12} = \frac{13}{12} = 1\frac{1}{12}$$

2. Solve linear simultaneous equations, e.g. solve the simultaneous equations
$$x + 3y = 9 \quad (1)$$
$$2x - y = 4 \quad (2)$$
Multiply (1) by 2: $2x + 6y = 18 \quad (3)$
Eqn (3) – Eqn (2): $7y = 14 \implies y = 2$
Substitute 2 for y in (1): $x + 6 = 9 \implies x = 3$

Skills check:

1. Write down the values of $\frac{12}{x}$ for $x = 1, 2, 3$ and 4.

2. Solve the simultaneous equations
$$0.6 + a + b = 1$$
$$3.2 + 2a + 4b = 4.6$$

5.1 Discrete random variables

> A **random variable** is a quantity that can take any value determined by the outcome of a random event.

Random variables can arise from probability experiments. For example, when you throw two dice,

X = the sum of the scores

is a random variable.
Similarly,

Y = product of the scores, and
Z = larger of the scores

are also random variables.

Random variables can arise from real-life observation. For example:

X = number of telephone calls arriving at a switchboard between 10 and 10:30 a.m.

When values of a variable have a probability attached, they form a **probability distribution**.

Similarly, when values have a frequency attached, they form a frequency distribution

> X is a **discrete random variable** if X takes values $x_1, x_2, x_3, \ldots,$
> and $P(X = x_i) = p_i$ where all $p_i \geq 0$ and $\sum p_i = 1$.

Note that this means X can only take distinct values, all probabilities of X must be non-negative and their sum is 1.

Example 1

X is a random variable with probability distribution given by:

x	-2	-1	0	1	2
$P(X = x)$	0.1	0.1	0.4	a	0.1

A probability distribution can be in the form of a table.

Find the value of a.

$\sum p_i = 1$

$1 - (0.1 + 0.1 + 0.4 + 0.1) = 0.3$
So $a = 0.3$

Example 2

If X = sum of the scores on two dice, the sample space for X will be:

X	1	2	3	4	5	6
1	2	3	4	5	6	7
2	3	4	5	6	7	8
3	4	5	6	7	8	9
4	5	6	7	8	9	10
5	6	7	8	9	10	11
6	7	8	9	10	11	12

Find the probability distribution for X.

The probability distribution for X is:

x	2	3	4	5	6	7	8	9	10	11	12
$P(X = x)$	$\dfrac{1}{36}$	$\dfrac{2}{36}$	$\dfrac{3}{36}$	$\dfrac{4}{36}$	$\dfrac{5}{36}$	$\dfrac{6}{36}$	$\dfrac{5}{36}$	$\dfrac{4}{36}$	$\dfrac{3}{36}$	$\dfrac{2}{36}$	$\dfrac{1}{36}$

Exercise 5.1

1. Find which of the following are discrete random variables, giving a reason for any which are not.

a)

x	1	2	3	4	5
$P(X = x)$	0.2	0.3	0.4	0.3	0.2

b)

x	−2	−1	0	1	2
$P(X = x)$	0.2	0.3	0.1	0.3	0.1

c)

x	5	7	10	15	20
$P(X = x)$	$\dfrac{1}{2}$	$\dfrac{1}{4}$	$\dfrac{1}{8}$	$\dfrac{1}{16}$	$\dfrac{1}{16}$

d)

x	1	2	3	4	5
$P(X = x)$	0.2	0.3	0.4	0.3	⁻0.2

2. A fair die is thrown. List the probability distribution for the following random variables:

a) X = score on the die

b) $Y = 2 \times$ score on the die

c) Z = square of the score on the die

d) $W = \begin{cases} 0 \text{ if the score on the die is a factor of } 6 \\ 1 \text{ otherwise} \end{cases}$

3. Two coins are tossed. If X = number of heads seen, list the probability distribution for X.

4. **a)**

x	1	2	3	4	5
$P(X = x)$	0.2	0.1	0.3	a	0.1

   i) Find a.   **ii)** Find $P(X \geq 2)$.

b)

x	-2	-1	0	1
$P(X = x)$	k	$2k$	$2k$	k

   i) Find k.   **ii)** Find $P(X \leq 0)$.

c)

x	5	6	8	9	12
$P(X = x)$	0.4	0.2	0.3	a	0.1

   i) Find a.   **ii)** Find $P(X \geq 9)$.

5.2 The probability function, $p(x)$

You can write the probability distribution of a discrete random variable as
    a **possibility space** – a table of values with their associated probabilities
    a **probability function** – a formula $p(x)$.
Here is an example of a probability function:
    $\Pr\{X = r\} = kr$ $r = 1, 2, 3, 4$
This means that the random variable X can take the values
1, 2, 3 or 4 with probabilities k, $2k$, $3k$, and $4k$.
You can then draw up a table showing all the probabilities explicitly:

> The total probability must be 1. When you add these up you get $10k$, so $k = 0.1$

r	1	2	3	4
Probability	0.1	0.2	0.3	0.4

> A probability function will often have an unknown constant in its expression, which can be found by setting the total probability equal to 1.

Example 3

$$\Pr\{X = r\} = \left(\frac{1}{6}\right)\left(\frac{5}{6}\right)^{r-1} \qquad r = 1, 2, 3, 4, \ldots$$

a) Write down the probabilities that $X = 1$, $X = 2$ and $X = 3$.
b) Find the probability that $X \geq 4$.

▶ Continued on the next page

a) $\Pr\{X = 1\} = \left(\dfrac{1}{6}\right)\left(\dfrac{5}{6}\right)^0 = \dfrac{1}{6}$

$\Pr\{X = 2\} = \left(\dfrac{1}{6}\right)\left(\dfrac{5}{6}\right)^1 = \dfrac{5}{36}$

$\Pr\{X = 3\} = \left(\dfrac{1}{6}\right)\left(\dfrac{5}{6}\right)^2 = \dfrac{25}{216}$

b) $\Pr\{X \geq 4\} = 1 - \Pr\{X = 1,\, 2,\, 3\} = 1 - \left(\dfrac{1}{6} + \dfrac{5}{36} + \dfrac{25}{216}\right) = \dfrac{125}{216}$

Example 4

The random variable X has probability function

$$\Pr\{X = k\} = \begin{cases} \dfrac{k-1}{36} & \text{for } k = 2,\, 3,\, 4,\, 5,\, 6,\, 7 \\[2mm] \dfrac{13-k}{36} & \text{for } k = 8,\, 9,\, 10,\, 11,\, 12 \end{cases}$$

This probability function describes the sum of the scores on two fair dice

Find: **a)** $P(X = 3)$ **b)** $P(X < 5)$.

a) $P(X = 3) = \dfrac{2}{36} = \dfrac{1}{18}$

b) $P(X < 5) = P(X = 2,\, 3 \text{ or } 4) = \dfrac{1}{36} + \dfrac{2}{36} + \dfrac{3}{36} = \dfrac{6}{36} = \dfrac{1}{6}$

Exercise 5.2

1. Which of the following could not be the probability function of a random variable? For those which could be, find the probability distribution.

   a) $\Pr\{X = r\} = kr$ for $r = 1,\, 2,\, 3,\, 4,\, 5$

   b) $\Pr\{X = r\} = \dfrac{k}{r}$ for $r = 1,\, 2,\, 3,\, 4$

   c) $\Pr\{X = r\} = \dfrac{k}{r-1}$ for $r = 1,\, 2,\, 3,\, 4$

   d) $\Pr\{X = r\} = \dfrac{k}{r+1}$ for $r = 1,\, 2,\, 3,\, 4$

2. For each of the following probability functions, list the probability distribution.

   a) $\Pr\{Z = z\} = \dfrac{5-z}{10}$ for $z = 1,\, 2,\, 3,\, 4$

   b) $\Pr\{Y = y\} = \dfrac{1}{5}$ for $y = 1,\, 2,\, 3,\, 4,\, 5$

   c) $\Pr\{W = w\} = \begin{cases} k(w-1) & \text{for } w = 2,\, 3,\, 4,\, 5,\, 6,\, 7 \\ k(13-w) & \text{for } w = 8,\, 9,\, 10,\, 11,\, 12 \end{cases}$

   d) $\Pr\{H = r\} = k(2r - 1)$ for $r = 1,\, 2,\, 3,\, 4$

3. X is a discrete random variable with probability function

$\Pr\{X = x\} = \dfrac{x}{15}$ for $x = 1, 2, 3, 4, 5$

A is the event $X \geq 3$ and B is the event $X < 4$.

Find

a) $P(A)$ **b)** $P(B)$ **c)** $P(A \cap B)$ **d)** $P(A \mid B)$ **e)** $P(B \mid A)$.

4. Z is a discrete random variable with probability function

$\Pr\{Z = z\} = \dfrac{k}{z}$ for $z = 1, 2, 3, 4$

a) Show that $k = 0.48$

b) Find

  i) $P(Z > 1)$ **ii)** $P(Z = 4 \mid Z > 1)$.

5. $\Pr\{Y = y\} = cy^2$ for $y = 1, 2, 3, 4$.

Find the value of c and hence find $\Pr\{Y < 3\}$.

5.3 Expectation of a discrete random variable

> The **mean** or **expected value** of a probability distribution is
> defined as $\mu = E(X) = \sum px$

This means that we:
multiply each value by its probability, and
sum over all possible values of the random
variable.

Remember in section 2.1 you met the mean of a frequency
distribution – and if you think of the probability p as being the
relative frequency of the value x then this definition parallels

the one met earlier: $\bar{x} = \dfrac{\sum xf}{n} = \sum \left\{ x \left(\dfrac{f}{n} \right) \right\} = \sum px$.

In section 5.4 a similar extension applies to the calculation of
the variance and standard deviation of a probability distribution.

Example 5

Members of a public library may borrow up to five books at any one time.

The number of books borrowed by a member on each visit is a random variable, X,
with the following probability distribution:

X	0	1	2	3	4	5
Probability	0.24	0.12	0.14	0.30	0.05	0.15

Find the mean of X.

$E(X) = (0 \times 0.24) + (1 \times 0.12) + (2 \times 0.14) + (3 \times 0.30) +$
$\qquad (4 \times 0.05) + (5 \times 0.15) = 2.25$

We multiply each value by its
probability, and take the sum of these.

The 'expected value' does not have to be a possible outcome

Example 6

X is a discrete random variable with probability distribution

x	1	2	3	4
p	$5k$	$2k$	k	$2k$

Show that $k = 0.1$ and calculate $E(X)$.

$5k + 2k + k + 2k = 10k = 1 \implies k = 0.1$

so

x	1	2	3	4
p	0.5	0.2	0.1	0.2

> The sum of all the probabilities will be equal to 1, and so we can use this to find a value for k.

$E(X) = (1 \times 0.5) + (2 \times 0.2) + (3 + 0.2) +$
$\quad (4 \times 0.2) = 2.3$

> Now that we have a value for each probability, we can find the mean by using the method above.

To find $E(X^2)$ – which will become useful in the next section – you can list the values of X^2 with the probability distribution of X.

So, for the library books from Example 5:

X^2	0	1	4	9	16	25
Probability	0.24	0.12	0.14	0.30	0.05	0.15

$E(X^2) = (0 \times 0.24) + (1 \times 0.12) + (4 \times 0.14) + (9 \times 0.30) + (16 \times 0.05) + (25 \times 0.15) = 7.93$

Example 7

X is a discrete random variable with probability distribution

x	1	2	3	4
p	a	0.3	b	0.2

$E(X) = 2.7$. Find the values of a and b.

$a + 0.3 + b + 0.2 = 1$ and $E(X) = a + 0.6 + 3b + 0.8 = 2.7$

so $\qquad a + b = 0.5$

$\qquad\qquad a + 3b = 1.3$

> Solve the two results as simultaneous equations

and $\qquad 2b = 0.8$

$\qquad\qquad b = 0.4$

$\qquad\qquad a = 0.1$

Exercise 5.3

1. For the following probability distributions, calculate the expectation of X.

a)

x	1	2	3	4	5
P(X = x)	0.2	0.1	0.2	0.3	0.2

b)

x	$^-2$	$^-1$	0	1	2
P(X = x)	0.2	0.3	0.1	0.3	0.1

c)

x	5	7	10	15	20
P(X = x)	$\dfrac{1}{2}$	$\dfrac{1}{4}$	$\dfrac{1}{8}$	$\dfrac{1}{16}$	$\dfrac{1}{16}$

2. For each of the following probability functions calculate the mean.

a) $\Pr\{Z = z\} = \dfrac{5-z}{10}$ for $z = 1, 2, 3, 4$

b) $\Pr\{Y = y\} = \dfrac{1}{5}$ for $y = 1, 2, 3, 4, 5$

3. a)

x	1	2	3	4	5
P(X = x)	0.3	0.2	0.1	a	0.1

   i) Find a.   **ii)** Find $E(X)$.   **iii)** Find $E(X^2)$.

b)

x	5	6	8	12
P(X = x)	k	2k	2k	k

   i) Find k.   **ii)** Find $E(X)$.   **iii)** Find $E(X^2)$.

4. Y is the larger score showing when two dice are thrown. Calculate $E(Y)$.

5. X is a probability distribution function and $E(X) = 3.4$.

x	1	2	3	4	5
P(X = x)	0.1	0.2	0.1	a	b

Find a and b.

6. X is a random variable and $E(X) = 7.4$.

x	4	6	7	10	11
P(X = x)	0.2	a	0.3	b	0.2

Find a and b. Find the probability that $X > E(X)$.

5.4 The variance of a discrete random variable

The **variance** of a probability distribution is defined as

$$\text{Var}(X) = E[\{X - E(X)\}^2].$$

The alternative version (which is easier to use in practice) is

$$\text{Var}(X) = E(X^2) - \{E(X)\}^2$$

> $E(X^2) = \sum px^2$
> That is, multiply all the x^2 values by their probabilities, and then find the total.
> Hence $\text{Var}(X) = \sum px^2 - \left(\sum px\right)^2$

An easy way to recall the alternative version is:

Find the mean of the squares, then subtract the square of the mean.

Example 8

X is a random variable with probability distribution

x	1	2	3	4
p	a	0.2	$3a$	0.2

a) Find the value of a.

b) Calculate $E(X)$ and $\text{Var}(X)$.

a) Since $\sum p = 1$,

$4a + 0.4 = 1$,

so $a = 0.15$

b) In table form:

x	p	px	px^2
1	0.15	0.15	0.15
2	0.2	0.4	0.8
3	0.45	1.35	4.05
4	0.2	0.8	3.2
		$\sum px = 2.7$	$\sum px^2 = 8.2$

> Var(X) = 0.91 provides a measure of how spread out the values of X are

$E(X) = \sum px = 2.7$

$\text{Var}(X) = E(X^2) - \{E(X)\}^2 = 8.2 - 2.7^2 = 0.91$

The **standard deviation** is the square root of the variance.

So in Example 7 it is $\sqrt{0.91} \approx 0.954$.

We often use Greek symbols as notation for the mean, variance and standard deviation:

$E(X) = \mu$

$\text{Var}(X) = \sigma^2$

Standard deviation of $X = \sigma$

> You have already met the symbols for variance and standard deviation, σ^2 and σ respectively, in Chapter 2.

Example 9

X is a random variable with probability distribution

x	1	2	3	4
p	a	0.2	b	0.2

If $E(X) = 2.6$, calculate

a) the values of a and b

b) $Var(X)$ and the standard deviation of X.

a) $a + b + 0.4 = 1$ ←   The sum of all the probabilities is equal to 1.

             $a + b = 0.6$

    $a + 0.4 + 3b + 0.8 = 2.6$ ←   We know that the mean is equal to the sum of the values multiplied by the probabilities.

           $a + 3b = 1.4$

so $2b = 0.8$

              $b = 0.4$ ←   By solving simultaneously.

              $a = 0.2$

b) In table form:

x	p	px	px^2
1	0.2	0.2	0.2
2	0.2	0.4	0.8
3	0.4	1.2	3.6
4	0.2	0.8	3.2
		$\sum px = 2.6$	$\sum px^2 = 7.8$

$Var(X) = E(X^2) - \{E(X)\}^2 = 7.8 - 2.6^2 = 1.04$

Standard deviation $= \sqrt{1.04} \approx 1.02$ ←   Variance is the square of the standard deviation.

Exercise 5.4

1. For each probability distribution, calculate $E(X)$ and $Var(X)$.

a)

x	5	6	7	8	9
$P(X = x)$	0.1	0.2	0.3	0.3	0.1

b)

x	‾2	‾1	0	1	2
$P(X = x)$	0.1	0.2	0.3	0.3	0.1

c)

x	1	2	3	4	5
$P(X = x)$	$\dfrac{1}{2}$	$\dfrac{1}{4}$	$\dfrac{1}{8}$	$\dfrac{1}{16}$	$\dfrac{1}{16}$

2. For each probability function, calculate the mean and variance.

a) $\Pr\{Z = z\} = \dfrac{6 - z}{15}$ $z = 1, 2, 3, 4, 5$

b) $\Pr\{Y = y\} = \dfrac{1}{6}$ $y = 1, 2, 3, 4, 5, 6$

c) $\Pr\{W = w\} = \begin{cases} k(w - 1) & \text{for } w = 2, 3, 4, 5, 6, 7 \\ k(13 - w) & \text{for } w = 8, 9, 10, 11, 12 \end{cases}$

3. a)

x	1	2	3	4	5
$P(X = x)$	0.2	0.2	0.2	a	0.1

   i) Find a.　　ii) Find $E(X)$ and $Var(X)$.

b)

x	3	4	5	6
$P(X = x)$	k	$2k$	$2k$	k

   i) Find k.　　ii) Find $E(X)$ and $Var(X)$.

4. Y is the smaller score when two dice are thrown.
Calculate $E(Y)$ and $Var(Y)$.

5. X is a random variable and $E(X) = 3.7$.

x	1	2	3	4	5
$P(X = x)$	0.1	0.2	0.1	a	b

Find a, b and $Var(X)$.

6. X is a random variable and $E(X) = 5.7$.

x	1	2	4	8	16
$P(X = x)$	0.1	a	0.3	b	0.1

Find a, b and $Var(X)$.

EXAM-STYLE QUESTIONS

1. The random variable X has probability function

$$\Pr\{X = x\} = \frac{k}{x} \qquad x = 1, 2, 3, 4$$

a) Show that $k = \frac{12}{25}$.

Find

b) $P(X < 3)$

c) $E(X)$.

2. The random variable X has probability function

$$\Pr\{X = x\} = \begin{cases} kx & \text{for } x = 3, 4, 5 \\ k(11-x) & \text{for } x = 6, 7, 8 \end{cases}$$

where k is a constant.

a) Show that $k = \frac{1}{24}$.

b) Find the exact value of $E(X)$.

c) Find $Var(X)$.

3. A discrete random variable X has the probability distribution shown in the table.

x	8	10	15
$P(X = x)$	0.4	a	$0.6 - a$

a) Given that $E(X) = 10.2$, find a.

b) Find $Var(X)$.

c) Find $P(X < \mu - \sigma)$.

4. An unbiased die is rolled. For each of the following, state whether the random variable is a discrete uniform distribution.

X = The number of times the die is rolled until a six shows.

Y = The score showing on the top face when the die is rolled.

$Z = 7$ = The score showing on the top face when the die is rolled.

A uniform discrete distribution takes all the values in its outcome space with equal probability.

5. A discrete random variable X has the probability distribution shown in the table below.

x	6	7	8	9
$F(x)$	0.1	0.2	0.3	0.4

Find

a) $P(X = 8)$ b) $E(X)$

c) $Var(X)$ d) $E(5 - 2X)$

e) $Var(5 - 2X)$.

6. a) A regular customer in a shop observes that the number of customers, X, in a shop when she enters has the following probability distribution.

No. of customers	0	1	2	3	4
Probability	0.1	0.25	0.3	0.25	0.1

i) Find the mean and standard deviation of X.

The same customer also observes that the average waiting time, Y, before being served is as follows:

No. of customers	0	1	2	3	4
Average wait (minutes)	0	2	5	8	11

ii) Find the mean waiting time.

b) The customer decides that, in future, if there are more than two customers waiting when she arrives, she will leave and return another day. On a return visit, she will stay no matter what the length of queue is.

i) What is the probability of her leaving the shop without waiting on her first visit?

ii) What is the probability that there are more customers in the shop when she returns for the second visit, following an occasion on which she has left without waiting?

7. The random variable X has probability distribution

x	7	8	9	10	11
$P(X = x)$	0.2	a	0.3	0.1	b

a) Given that $E(X) = 9.05$, write down two equations involving a and b.

Find

b) the value of a and the value of b

c) $Var(X)$.

8. The random variable X has probability function
$$Pr(X = x) = \frac{11 - 2x}{25} \qquad x = 1, 2, 3, 4, 5$$

a) Construct a table giving the probability distribution of X.

Find

b) $P(2 < X < 5)$

c) $E(X)$

d) $Var(X)$.

9. The random variable X has the cumulative probability distribution

x	10	12	15	16	18	20	24	25
$P(X \leq x)$	0.1	0.3	0.35	0.4	0.6	0.7	0.9	1

Find

a) the probability distribution of X.

b) $E(X)$

c) $Var(X)$.

10. The discrete random variable X has probability function
$$Pr\{X = x\} = kx^2 \qquad x = 1, 2, 3, 4,$$
where k is a positive constant.

a) Show that $k = \dfrac{1}{30}$.

b) Find $E(X)$ and show that $E(X^2) = 11.8$.

c) Find $Var(X)$.

11. An independent financial adviser recommends investments in seven traded options. The number from which the client makes a profit can be modelled by the discrete random variable X with probability function
$$Pr\{X = x\} = kx \qquad x = 0, 1, 2, 3, 4, 5, 6, 7,$$
where k is a constant.

a) Find the value of k.

b) Find $E(X)$ and $Var(X)$.

The total investment is $4000 and the return on each successful option is $1500.

> Draw up a new probability distribution showing the profit and loss rather than the number of successful options.

c) Find the probability that the client makes a loss overall.

d) Find the mean and variance of the profit that the client makes.

12. Sanket plays a game using a biased die which is twice as likely to land on an even number as on an odd number. The probabilities for the three even numbers are all equal and the probabilities for the three odd numbers are all equal.

(i) Find the probability of throwing an odd number with this die. [2]

Sanket throws the die once and calculates his score by the following method.

- If the number thrown is 3 or less he multiplies the number thrown by 3 and adds 1.
- If the number thrown is more than 3 he multiplies the number thrown by 2 and subtracts 4.

The random variable X is Sanket's score.

(ii) Show that $P(X = 8) = \dfrac{2}{9}$. [2]

The table shows the probability distribution of X.

x	4	6	7	8	10
$P(X = x)$	$\dfrac{3}{9}$	$\dfrac{1}{9}$	$\dfrac{2}{9}$	$\dfrac{2}{9}$	$\dfrac{1}{9}$

(iii) Given that $E(X) = \dfrac{58}{9}$, find $Var(X)$. [2]

Sanket throws the die twice.

(iv) Find the probability that the total of the scores on the two throws is 16. [2]

(v) Given that the total of the scores on the two throws is 16, find the probability that the score on the first throw was 6. [3]

Cambridge International AS and A Level Mathematics 9709, Paper 61 Q7 October/November 2010

13. A fair die has one face numbered 1, one face numbered 3, two faces numbered 5 and two faces numbered 6.

(i) Find the probability of obtaining at least 7 odd numbers in 8 throws of the die. [4]

The die is thrown twice. Let X be the sum of the two scores. The following table shows the possible values of X.

		Second throw				
		1	3	5	5	6	6
First throw	1	2	4	6	6	7	7
	3	4	6	8	8	9	9
	5	6	8	10	10	11	11
	5	6	8	10	10	11	11
	6	7	9	11	11	12	12
	6	7	9	11	11	12	12

(ii) Draw up a table showing the probability distribution of X. [3]

(iii) Calculate $E(X)$. [2]

(iv) Find the probability that X is greater than $E(X)$. [2]

Cambridge International AS and A Level Mathematics 9709, Paper 6 Q7 October/November 2008

14. In a competition, people pay $1 to throw a ball at a target. If they hit the target on the first throw they receive $5. If they hit it on the second or third throw they receive $3, and if they hit it on the fourth or fifth throw they receive $1. People stop throwing after the first hit, or after 5 throws if no hit is made. Mario has a constant probability of $\dfrac{1}{5}$ of hitting the target on any throw, independently of the results of other throws.

(i) Mario misses with his first and second throws and hits the target with his third throw. State how much profit he has made. [1]

(ii) Show that the probability that Mario's profit is $0 is 0.184, correct to 3 significant figures. [2]

(iii) Draw up a probability distribution table for Mario's profit. [3]

(iv) Calculate his expected profit. [2]

Cambridge International AS and A Level Mathematics, 9709 Paper 6 Q6 October/November 2005

15. The probability distribution of the discrete random variable X is shown in the table below.

x	$^-3$	$^-1$	0	4
$P(X = x)$	a	b	0.15	0.4

Given that $E(X) = 0.75$, find the values of a and b. [4]

Cambridge International AS and A Level Mathematics 9709, Paper 61 Q1 May/June 2010

16. The numbers of rides taken by two students, Fei and Graeme, at a fairground are shown in the following table.

	Roller coaster	Water slide	Revolving drum
Fei	4	2	0
Graeme	1	3	6

(i) The mean cost of Fei's rides is $2.50 and the standard deviation of the costs of Fei's rides is $0.
Explain how you can tell that the roller coaster and the water slide each cost $2.50 per ride. [2]

(ii) The mean cost of Graeme's rides is $3.76. Find the standard deviation of the costs of Graeme's rides. [5]

Cambridge International AS and A Level Mathematics 9709, Paper 61 Q4 May/June 2010

17. A team of 4 is to be randomly chosen from 3 boys and 5 girls. The random variable X is the number of girls in the team.

i) Draw up a probability distribution table for X. [4]

ii) Given that $E(X) = \dfrac{5}{2}$, calculate Var(X). [2]

Cambridge International AS and A Level, Paper 61 Q3 October/November 2011

18. Ashok has 3 green pens and 7 red pens. His friend Rod takes 3 of these pens at random, without replacement. Draw up a probability distribution table for the number of green pens Rod takes. [4]

Cambridge International AS and A Level Mathematics 9709, Paper 61 Q1 October/November 2012

Chapter summary

- A random variable is a quantity that can take any value determined by the outcome of a random event.

- X is a **discrete random variable** if X takes values $x_1, x_2, x_3, \ldots$, and $P(X = x_i) = p_i$ where all $p_i \geq 0$ and $\sum_i p_i = 1$.

- A probability function will often have an unknown constant in its expression, which can be found by setting the total probability equal to 1.

- The mean or expected value of a probability distribution is defined as $\mu = \mathrm{E}(X) = \sum px$

- The variance of a probability distribution is defined as $\mathrm{Var}(X) = \mathrm{E}[\{X - \mathrm{E}(X)\}^2]$.
  The alternative version (which is easier to use in practice) is
  $\mathrm{Var}(X) = \mathrm{E}(X^2) - \{\mathrm{E}(X)\}^2$

- The standard deviation is the square root of the variance.

Campanology is the art of bell ringing – usually church bells. These bells are very heavy and each bell requires a person to swing it. The inertia of the bells is so great that the ringers have only a limited ability to advance or slow down their bell in the sequence they are rung in. In most cases the number of bells is limited (6 is common) so the number of available notes is limited, as well as the ability to change the sequence. The study of permutations has been prominent in the evolution of bell ringing since the seventeenth century. An 'extent' is a sequence in which all possible permutations of the n bells in the tower have been rung exactly once each. With six bells an extent will take around half an hour, but the time taken for an extent grows very quickly as the number of bells increases – with 8 bells it would take nearly a full day and night. By the time there are 12 bells it would be around 30 years!

Objectives

After studying this chapter you should be able to:

- Understand the terms permutations and combinations, and solve simple problems involving selections.
- Solve problems about arrangements of objects in a line, including those involving:
  - repetition (e.g. the number of ways of arranging the letters of the word 'NEEDLESS'),
  - restriction (e.g. the number of ways several people can stand in a line if 2 particular people must — or must not — stand next to each other).
- Evaluate probabilities by calculation using permutations and combinations.

Before you start
You should know how to:

1. List the possible outcomes in a given situation, e.g.

   Three coins are tossed. List the possible outcomes.

   HHH, HHT, HTH, HTT, THH, THT, TTH, TTT

Skills check:

1. Two coins are tossed. List the possible outcomes.

2. A coin is tossed and a die is thrown. List the possible outcomes.

6.1 Permutations of *n* distinct objects in a straight line

If we want to know how many ways two objects can be arranged, it is easy to see that there are two possible orders – in the form AB and BA.

If we have three objects, we can argue that there are three ways to place the first object. We then have two objects left to place in order, and we already know that can be done in two ways, which gives us a total of six possible orders.

We can list these six permutations easily, however when we have a larger number of orders, it becomes more time consuming. The reasoning used above (known as inductive reasoning) is very powerful and much quicker – we will use it shortly to generalise these results to provide an expression for the number of orders of *n* distinct objects.

If we apply the same reasoning to the first example, where we only had two objects, we can say that we had two ways to put in the first letter of the sequence and then only one way to put in the remaining letter so the number of permutations of two distinct objects is $2 \times 1 = 2$.

A	B	C
A	C	B
B	A	C
B	C	A
C	A	B
C	B	A

For each of the three starting letters, the two remaining letters are first in alphabetical order and then in reverse order. The number of permutations of three distinct objects is then $3 \times (2 \times 1) = 6$.

Inductive reasoning now lets us argue that the number of permutations of four distinct objects in a line will be $4 \times 6 = 24$, as we can put the first object in four ways, and we know the other three can then be arranged in six ways in each case.

> This is why you multiply 4 by 6. So it is 4 x (3 x (2 x 1)) = 24

For small *n*, writing this out in full is simple enough, but it takes a long time so mathematicians have introduced a notation called factorials to shorten it:

> $n! = n(n-1)(n-2) \ldots, \ldots 3.2.1$ where *n* is a positive integer.
> For reasons which will become clear shortly, it is convenient to define $0! = 1$.

> We read '*n*!' as '*n* factorial'.

> The number of ways of arranging *n* distinct objects in a straight line is *n*!

You should get to know the first few (1, 2, 6, 24) and your calculator will give you the others. However, because the numbers grow so quickly, the calculator will display them in standard form once it does not have enough room to display the number in full – for example a calculator normally shows 13! as 6 227 020 800 and 14! as $8.717\,829\,12 \times 10^{10}$.

Note that if you divide a large factorial by a smaller one, you can cancel a lot of terms:

$6! = 6 \times 5 \times 4 \times 3 \times 2 \times 1$

$ = 6 \times 5 \times 4!$, so

$\dfrac{6!}{4!} = 6 \times 5 = 30.$

You will find this useful later in the chapter.

Example 1

a) In how many ways can five people sit on a row of five chairs?

b) A race has four runners. In how many orders can they finish if they all finish at different times?

c) How many ways can the letters of the word CAMBRIDGE be arranged?

a) 5! = 120

b) 4! = 24 ← It does not matter if it is people or letters to be arranged in order – for all of these it is arranging n distinct 'objects' in order in a line.

c) 9! = 362 880

Exercise 6.1

1. Calculate

   a) 6!   **b)** 3!   **c)** 12!

2. Calculate

   a) $\dfrac{7!}{6!}$ **b)** $\dfrac{5!}{3!}$ **c)** $\dfrac{8!}{8!}$

3. How many ways can the letters of the word YEAR be arranged?

4. There are eight athletes in a race. Assuming they all finish the race and they all finish at different times, how many results of the race are possible?

5. A set of eight cards each have a different colour. The cards are shuffled randomly and then laid out in a line. How many different arrangements are possible?

6. How many different numbers can be made using each of the digits 1 to 5 exactly once each?

7. How many ways can the letters of the word MATHS be arranged?

8. In how many ways can ten people sit on a row of ten chairs?

6.2 Permutations of *k* objects from *n* distinct objects in a straight line

While in some contexts the full order of objects may be of interest, often not all of the objects will be needed. For example in competitions there may be medals for the first three places and so the ordering is only important for the first few places.

Example 2

In a race with eight competitors how many ways can the gold, silver and bronze medals be awarded?

$8 \times 7 \times 6 = 336$ ⟵ Gold can be any one of 8, then silver any one of the remaining 7 and bronze any one of 6.

We talk about **permutations of 3 from 8 distinct items** to describe the above, and the general formula then is:

> The number of permutations of *k* objects from *n* distinct objects in a straight line is given by $^nP_k = \dfrac{n!}{(n-k)!}$.

Calculators will have a function key combination for ^nP_k or you can calculate it using the factorials (or by cancelling if *k* is close to *n*).

Mathematicians define 0! to be 1.

This is very similar to defining x^0 to be equal to 1 for any $x \neq 0$, which means that the law of indices $\dfrac{x^m}{x^n} = x^{m-n}$ holds when $m = n$ where the left hand side equals 1 because the top and bottom of the fraction are equal (and not 0).

The number of permutations of 8 objects from 8 was 8!, but using the above formula it would be $^8P_8 = \dfrac{8!}{0!}$ which will only equal 8! if 0! takes the value 1.

Exercise 6.2

1. Calculate, without using a calculator
   a) $\dfrac{8!}{7!}$
   b) $\dfrac{15!}{13!}$
   c) $\dfrac{120!}{119!}$
   d) $\dfrac{6!}{3!}$

2. Use your calculator to work out: a) $\dfrac{18!}{12!}$
   b) $^{15}P_{12}$

3. How many different five letter arrangements can be made using the letters in CAMBRIDGE.

4. How many different four digit numbers can be made using the digits 1 to 9 at most once each?

5. Ten men are in a doctor's waiting room which has a row of eight chairs. How many different arrangements are there of men sitting in the chairs?

6. There are eight athletes in a race. Assuming they all finish the race and each athlete finishes at a different time, how many results of the top five finishers are possible?

7. A set of eight cards each have a different colour, They are shuffled randomly and then three of them are laid out in a line. How many different arrangements are possible?

8. A candidate in an examination has to answer four questions from a choice of six, and can answer them in any order. How many different orders can the candidate provide their answers in?

6.3 Allowing constraints on permutations (for *n* distinct objects)

Often there are constraints placed on the permutations requiring something to happen (or not to happen).

Example 3

There are six people including a husband and wife, and a row of six chairs. In how many ways can they sit if the husband and wife are (i) to sit together and (ii) not to sit together?

i) $2 \times 5! = 240$

Think of the married couple as one 'object' – so there are five objects to be placed in order. But for each position the pair can appear in, there are two orders for the husband and wife.

ii) $6! = 720$

There are 6! ways the six people can sit with no constraints.

$2 \times 5! = 240$

$740 - 240 = 480$

Of these 6! permutations, $2 \times 5!$ had the married couple sitting together.

So in the rest of the cases they are not sitting together.

You would not normally be asked to calculate both of these – but if you were asked part (ii) you would be advised to find the number of times that they are together because trying to find a way of calculating how often they are apart directly is very difficult.

> The difficult part of these problems is identifying how to place the objects or choose the objects to satisfy the given constraints.

Try to think of a physical process of assigning objects to positions – as described in the white comment boxes above – before looking at the numbers.

Example 4

Eight books are to be arranged on a bookshelf. Three of the books are by the same author. In how many ways can the books be arranged on the bookshelf if the books by the same author are to stay together?

$3! \times 6! = 6 \times 720 = 4320$

Note that you multiply 6 by 720 because you get 6 ways for every one of the 720 orders you had already identified.

First treat the 3 books by the same author as one 'object' – and arrange the 6 objects on the bookshelf. Then for each one of those arrangements, there are $3! = 6$ orders in which the 3 books by the same author can be arranged.

Example 5

How many ways can the letters of the word NEPAL be arranged so that there is not a vowel at either end?

$3 \times 2 \times 3! = 36$ ways.

We need to put a consonant in the first place (3 ways); then a consonant in the last place (2 ways) and then fill the remaining 3 spaces with the 3 remaining letters (3! ways).

Example 6

How many five digit even numbers greater than 40 000 can be made using the digits 1 to 5 exactly once each?

$1 \times 1 \times 3! + 1 \times 2 \times 3!$
$= 6 + 12 = 18.$

Here we need to consider two cases separately. First we need to consider starting with 4 and with 5 separately, because when it starts with 4 then it must end with 2 to be even, so there are 3! ways to put in 1, 3 and 5. If it starts with 5 then there is a choice of 2 numbers (2 or 4) to end, and then you have to put in the other 3 digits.

Exercise 6.3

1. A group of six people includes two men. In how many ways can they be seated in a row of six chairs so that the two men sit beside one another.

2. A group of eight people includes two married couples. In how many ways can they be seated on a row of eight chairs if
   a) there are no restrictions
   b) the married couples each sit beside their partner.

3. Ten books have to be arranged on a bookshelf. Three of the books are poetry. In how many ways can they be arranged on the bookshelf if

   a) the three poetry books have to be together

   b) the poetry books are volumes I, II and III of a collection and must appear in order.

4. How many five letter arrangements which start with a consonant can be made from the letters of the word MATHS?

5. How many five letter arrangements which end with a vowel can be made from the letters of the word MATHS?

6. How many odd five digit numbers greater than 30 000 can be made from the digits 1 to 5?

7. The five digits 1, 2, 3, 4, 5 can be arranged to give many different five digit numbers.

   a) how many different five digit numbers are odd?

   b) how many of them start with a 4?

8. Izzy has 14 different CDs, of which three are film soundtracks, five are jazz and six are classical. How many different arrangements are there of the 14 CDs if

   a) the classical CDs are kept next to one another

   b) the CDs of each type are kept together?

9. A singer has seven songs in the set she is to perform. Three of them are ballads which have to be performed all together at the end of the set. How many different arrangements of songs can she have?

6.4 Permutations when some objects are not distinguishable

If we want to arrange the letters of MATHS, all the letters are different and we know there are 5! ways to do this. How many ways are there to arrange the letters of STATS where the letters are not all different?

This is much easier than it appears if we reason as follows: suppose the letters are all distinguishable (think of them as $S_1T_1AT_2S_2$. Then we know that there are 5! permutations possible. Of those, the Ts appear in the same places as T_1 followed by T_2 and also as T_2 followed by T_1. So we can divide the number of orders (5!) by 2! – the number of orders of the two identical Ts. We can then do the same where the two S letters appear, and divide again by 2!. So there will be $\frac{5!}{2!2!} = 30$ ways to arrange the letters of STATS.

> The number of permutations of n objects, of which p are identical to one another, q of the remainder are identical to one another, and so on (with $p + q + ... = n$, and the values of p, q etc. can be 1) is given by $\frac{n!}{p!q!r!s!...}$.

Example 7

Find the number of distinct permutations of the letters of the words

a) OXFORD

b) MISSISSIPPI

....

a) $\dfrac{6!}{2!} = 360$ ⟵ Here there are six letters with two the same.

b) $\dfrac{11!}{4!4!2!} = 34650$ ⟵ Here there are 11 letters altogether with 4, 4, 2 and 1 of 1, S, P and M respectively.

Exercise 6.4

1. Find the number of distinct permutations of the letters of the words
   a) TRINIDAD
   b) TOBAGO.

2. Find the number of distinct permutations of the letters of the words
   a) ASSIST
   b) HINDER.

3. The digits 3, 3, 3, 5, 6, 6, 7, and 7 are to be used to make an eight digit code. How many distinct codes are there?

4. Another eight digit code consists of three '1's and five '0's. How many distinct codes are there?

5. A 16 digit code consists of two blocks of eight. The first contains two '1's and six '0's and the second contains four '1's and four '0's. How many distinct codes are there?

6. Another 16 digit code contains six '1's and ten '0's. How many distinct codes are there?

7. There are ten spaces to park bicycles at a school. There are six bicycles on a particular day. How many ways are there to put these six bicycles into the ten spaces if
   a) there are no restrictions
   b) the four empty spaces are to be beside one another.

6.5 Combinations

In many cases the order of selection matters (as the previous sections have dealt with), but in others the order of selection does not matter. When the order of selection does not matter we refer to the number of **combinations** – or the number of ways to choose k objects out of n.

> The number of ways to choose k from n distinct objects is $^nC_k = \begin{pmatrix} n \\ k \end{pmatrix} = \dfrac{n!}{k!(n-k)!}$

You can see why this will be the case by considering the permutations of k from n distinct objects, which was $^nP_k = \dfrac{n!}{(n-k)!}$, but now each group

of k objects chosen will appear in $k!$ different orders, giving $\dfrac{\frac{n!}{(n-k)!}}{k!} = \dfrac{n!}{k!(n-k)!}$

as the number of choices of k objects out of n.

Example 8

A committee of five people is to be chosen from eight volunteers. How many different committees are possible.

$$\begin{pmatrix} 8 \\ 5 \end{pmatrix} = \frac{8!}{5!3!} = 56$$ ← Choosing five out of eight with no restrictions is always $\begin{pmatrix} 8 \\ 5 \end{pmatrix}$ ways.

Example 9

A committee of five people is to be chosen from seven men and six women. How many different committees are possible which have three men and two women.

$$\begin{pmatrix} 7 \\ 3 \end{pmatrix} \times \begin{pmatrix} 6 \\ 2 \end{pmatrix} = 35 \times 15 = 525$$ ← We have to choose three out of seven men and two out of six women.

Note: These have to be multiplied because each group of men can be matched with any of the 15 possible combinations of women on the committee.

Example 10

A committee of five people is to be chosen from seven men and six women. How many different committees will have a majority of men?

$$\binom{7}{3} \times \binom{6}{2} + \binom{7}{4} \times \binom{6}{1} + \binom{7}{5} \times \binom{6}{0}$$

$$= (35 \times 15) + 3(5 \times 6) + (21 \times 1)$$

$$= 525 + 210 + 21 = 756$$

As with example 6, we need to consider different committee make ups separately. 3, 4 or 5 men will give a majority. Each of those cases is calculated in a similar fashion to example 9.

Example 11

A committee of five people is to be chosen from eight volunteers, including a husband and wife. How many different committees are possible which do not include both husband and wife.

$$\binom{8}{5} - \binom{6}{3} = 56 - 20 = 36$$

We know choosing five out of eight with no restrictions is $\binom{8}{5}$, and if both are included then there are six others from which to choose the remaining three committee members.

Example 12

A committee of five is to be chosen from eight volunteers, including a husband and wife. If the committee members are chosen randomly, what is the probability that the committee includes both husband and wife?

$$\frac{20}{56} = \frac{5}{14}$$

In Example 11 we saw that there were 56 possible committees altogether, of which 20 included both husband and wife.

The probability that a selection satisfying a particular condition is chosen is the ratio of the number of possibilities satisfying the condition to the total number of possibilities.

Exercise 6.5

1. In a class of 30 students, four are chosen randomly to attend a presentation by a guest speaker. How many different selections are possible?

Note: The term 'selection' implies that order is not important and the term 'arrangement' implies the order is important.

2. Five cards are to be selected at random from a set of 20 different cards. In how many ways can this be done?

3. **a)** Show (algebraically) that choosing r objects out of n distinct objects gives the same number of selections as choosing $(n–r)$ objects out of n.

   b) Explain why (in practical terms) there must be the same number of possible selections in these two cases.

4. Four letters are to be selected at random from the English alphabet. How many possible selections are there if

   a) there are no restrictions

   b) there has to be at least one vowel.

5. Drug testing authorities arrive at a cycling race to test 20% of the riders. There are 25 riders, and those to be tested are selected randomly. How many possible combinations of riders are there who could be tested?

6. A new strategy is to be adopted by the drug testing authorities so they always test the winner and runner up in a race and then select the rest of the 20% sample randomly. In the race with 25 riders, how many possible combinations of riders will be tested? Give a reason why the new strategy might have been suggested.

A final observation: a lot of contexts involve a mixture of permutations and combinations. The key is to have a clear idea of how the process is to be carried out, including whether the order is important in any part of the process. It is also worth noting that there is often more than one strategy for identifying all the possibilities – but if they are correct then they will give the same number of possibilities at the end.

Example 13

Find the number of ways of arranging five men and three women in a row so there are not two women standing together anywhere in the row.

If you arrange the five men in order (5! ways to do this) and leave a space between each pair of men, with spaces at each end, there are then 6 spaces in which a woman can be put and satisfy the required condition.

Strategy 1:

Choose 3 out of 6 spaces $\left(\binom{6}{3} = 20 \text{ ways}\right)$, and then arrange the 3 women in those 3 spaces (3! ways to do this). This gives a total of $5! \times 20 \times 6$ ways to arrange the group with no women standing together.

Strategy 2:

Put the first woman in one of the six spaces, then the second woman in one of the remaining 5 spaces and the last woman can go in any one of 4 spaces, which also gives $5! \times 6 \times 5 \times 4$ ways.

6.6 Evaluate probabilities by calculation using permutations or combinations

In this chapter so far we have concentrated on calculating the number of ways some event can happen. You will often be asked to use these principles to calculate a probability of seeing a certain event, where it can be calculated as the ratio of the number of outcomes that satisfy that event to the total number of outcomes.

Example 14

Five men and three women are arranged randomly in a row. What is the probability that there are not two women standing together anywhere in the row?

This is the probability version of example 13

In example 13 the number of ways this restricted condition could occur was $5! \times 20 \times 6$.

Putting 8 people into a row with no restrictions can be done in 8! ways so the required probability is the ratio of these two counts:

$P \text{ (not two women standing together anywhere in the row)} = \dfrac{5! \times 20 \times 6}{8!} = \dfrac{5}{14}$

Exercise 6.6

1. Three letters are selected at random from the word STUDY. What is the probability that the selection does not contain any vowels?

2. Four letters are selected at random from the letters in the word CAMBRIDGE. What is the probability that the selection does not contain any vowels?

3. A team of 5 is to be selected from a group of 11 people in which there are 2 girls, 4 boys and 5 adults. If the team members are selected randomly find the probability that

   a) both girls are in the team

   b) there are no adults in the team

   c) there are at least two adults in the team.

4. A class has 14 boys and 16 girls and 4 students are selected at random. What is the probability that the selection has exactly two boys in it?

5. A standard pack of playing cards is well shuffled and dealt out equally to four players. What is the probability that a particular player has no diamonds?

6. A group of 8 people is made up of 4 husband and wife pairs. If they stand randomly in a line what is the probability that each husband will be standing next to his own wife?

Summary exercise 6

1. A soccer team is taking 16 people to a match in four cars. The owners of the cars drive their own cars, and each take three other players as passengers. In how many ways can this be done?

2. A group of ten people consists of five married couples. There are arranged in a line, and randomly allocated a position. Find the probability that each wife is standing beside her husband.

EXAM-STYLE QUESTIONS

3. A three digit number is made by choosing three different digits from 2, 3, 5, 7, and 8. What is the probability that the number chosen will be divisible by 3?

4. Six cards numbered 2, 4, 6, 7, 9, 11 are arranged randomly in a line. What is the probability that a card differs from a neighbour by only one?

5. Issam has 11 different CDs, of which 6 are pop music, 3 are jazz and 2 are clasical.

   (i) How many different arrangements of all 11 CDs on a shelf are there if the jazz CDs are all next to each other? [3]

   (ii) Issam makes a selection of 2 pop music CDs, 2 jazz CDs and 1 classical CD. How my different possible selections can be made? [3]

   Cambridge International AS and A level Mathematics 9709, Paper 6 Q3 May/June 2008

6. **a)** In a sweet shop 5 identical packets of toffees, 4 identical packets of fruit gums and 9 identical packets of chocolates are arranged in a line on a shelf. Find the number of different arrangements of the packets that are possible if the packets of chocolates are kept together. [2]

b) Jessica buys 8 different packets of biscuits. She then chooses 4 of these packets.

    i) How many different choices are possible if the order in which Jessica chooses the 4 packets is taken into account? [2]

The 8 packets include 1 packet of chocolate biscuits and 1 packet of custard creams.

    ii) How many different choices are possible if the order in which Jessica chooses the 4 packets is taken into account and the packet of chocolate biscuits and the packet of custard creams are both chosen? [3]

c) 9 different fruit pies are to be divided between 3 people so that each person gets an odd number of pies. Find the number of ways this can be done. [5]

Cambridge International AS and A level Mathematics 9709, Paper 61, Q7, October/November 2012

7. a) Seven friends together with their respective partners all meet up for a meal. To commemorate the occasion they arrange for a photograph to be taken of all 14 of them standing in a line.

    i) How many different arrangements are there if each friend is standing next to his or her partner? [3]

    ii) How may different arrangements are there if the 7 friends all stand together and the 7 partners all stand together? [2]

b) A group of 9 people consists of 2 boys, 3 girls and 4 adults. In how many ways can a team of 4 be chosen if

    i) both boys are in the team, [1]

    ii) the adults are either all in the team or all not in the team, [2]

    iii) at least 2 girls are in the team? [2]

Cambridge International AS and A level Mathematics 9709, Paper 61 Q7 May/June 2012

8. A committee of 5 people is to be chosen from 6 men and 4 women. In how many ways can this be done

    i) if there must be 3 men and 2 women on the committee, [2]

    ii) if there must be more men than women on the committee, [3]

    iii) if there must be 3 men and 2 women, and one particular woman refuses to be on the committee with one particular man? [3]

Cambridge International AS and A level Mathematics 9709, Paper 6 Q5 May/June 2003

9. The work ARGENTINA includes the four consonants R, G, N, T and the three vowels A, E, I.

    i) Find the number of different arrangements using all nine letters. [2]

    ii) How many of these arrangements have a consonant at the beginning, then a vowel, then another consonant, and so on alternately? [3]

Cambridge International AS and A level Mathematics 9709, Paper 6 Q1 October/November 2004

10. A staff car park at a school has 13 parking spaces in a row. There are 9 cars to be parked.

    i) How many different arrangements are there for parking the 9 cars and leaving 4 empty spaces? [2]

    ii) How many different arrangements are there if the 4 empty spaces are next to each other? [3]

    iii) If the parking is random, find the probability that there will **not** be 4 empty spaces next to each other. [2]

Cambridge International AS and A level Mathematics 9709, Paper 6 Q3 October/November 2005

Chapter summary

- The number of ways of arranging n distinct objects in a straight line is $n! = n(n-1)(n-2)...,$ ..., 3.2.1 where n is a positive integer. $n!$ is read as n factorial.

- $0! = 1$.

- The number of permutations of k objects from n distinct objects in a straight line is given by $^nP_k = \dfrac{n!}{(n-k)!}$.

- The number of permutations of n objects, of which p are identical to one another, q of the remainder are identical to one another, and so on (with $p + q + ... = n$, and the values of p, q etc. can be 1) is given by $\dfrac{n!}{p!q!r!s!......}$.

- The number of ways to choose k from n distinct objects is $^nC_k = \dbinom{n}{k} = \dfrac{n!}{k!(n-k)!}$

- Where there are constraints on the order or composition or a group, you need to identify how to place the objects or choose the objects to satisfy the given constraints.

Review exercise B

1. A fair die has six faces numbered 1, 1, 2, 2, 2 and 3. The die is rolled twice and the number showing on the uppermost face is recorded each time.

   Find the probability that the sum of the two numbers recorded is not more than 4.

2. Events A and B are defined in the sample space S. The events A and B are independent.

   Given that $P(A) = 0.2$, and $P(A \cup B) = 0.6$, find

   a) $P(B)$

   b) $P(A|B)$,

   A and C are mutually exclusive and $P(C) = 0.6$.

   c) Find $P(A \cup C)$

   d) Find the maximum and minimum possible values for $P(B \cup C)$.

3. In a school, 50% of the pupils taking Mathematics also took Chemistry. While 30% took Biology as well as Mathematics, and 10% took both Chemistry and Biology. One of the pupils taking Mathematics is chosen at random.

   a) Find the probability that the pupil took neither Chemistry nor Biology.

   b) Given that the pupil took exactly one of Chemistry and Biology, find the probability it was Chemistry.

4. Data about part-time jobs for males and females studying at a college are shown in the table.

	Works part-time	Not working
Male	263	165
Female	223	210

   A student at the college is chosen at random. Let F be the event that the student is female and W be the event that the student works part-time.

   i) Find $P(F)$

   ii) Find $P(F$ and $W)$

   iii) Are F and W independent events? Justify your answer.

   iv) Given that the student works part-time, find the probability that the student is male.

5. A discrete random variable X has the probability function shown in the table below.

x	8	12	20
$P(X = x)$	0.2	a	$0.8-a$

   a) Given that $E(X) = 15.2$, find a.

   b) Find Var(X).

   c) Find $P(X < \mu - \sigma)$ where $\mu = E(X)$ and $\sigma^2 = \text{Var}(X)$

6. The random variable X has probability distribution

x	3	4	5	6	7
$P(X = x)$	0.1	a	0.3	0.2	b

   a) Given that $E(X) = 5.5$, write down two equations involving a and b.

   Find

   b) the value of a and the value of b,

   c) Var(X),

7. The random variable X has probability function

   $$P(X = x) = \frac{(1+2x)}{24} \qquad x = 1, 2, 3, 4$$

   a) Construct a table giving the probability distribution of X.

   Find

   b) $P(1 < X < 4)$ c) $E(X)$ d) Var(X)

8. The discrete random variable X has probability function $P(X = x) = kx^3$ for $x = 1, 2, 3$ where k is a positive constant.

   a) Show that $k = \dfrac{1}{36}$.

   b) Find $E(X)$ and show that $E(X^2) = \dfrac{98}{36}$.

   c) Find $Var(X)$.

9. How many different five digit numbers can be made using the digits 0 to 9 at most once each (0 cannot be used as the first digit).

10. Twelve books have to be arranged on a bookshelf. Three of the books are by the same author. In how many ways can they be arranged on the bookshelf if

   a) the three books by the same author have to be together

   b) the books by the same author are actually parts I, II and III of a trilogy and must appear in order.

11. Five letters are to be selected at random from the English alphabet. How many possible selections are there if

   a) there are no restrictions

   b) there has to be at least two consonants?

12. The people living in 3 houses are classified as children (C), parents (P) or grandparents (G). The numbers living in each house are shown in the table below.

House number 1	House number 2	House number 3
4C, 1P, 2G	2C, 2P, 3G	1C, 1G

   i) All the people in all 3 houses meet for a party. One person at the party is chosen at random. Calculate the probability of choosing a grandparent. [2]

   ii) A house is chosen at random. Then a person in that house is chosen at random. Using a tree diagram, or otherwise, calculate the probability that the person chosen is a grandparent. [3]

   iii) Given that the person chosen by the method in part (ii) is a grandparent, calculate the probability that there is also a parent living in the house. [4]

   Cambridge International AS and A Level Mathematics 9709, Paper 6 Q6 May/June 2003

13. Two fair dice are thrown.

   i) Event A is 'the scores differ by 3 or more'. Find the probability of event A. [3]

   ii) Event B is 'the product of the scores is greater than 8'. Find the probability of event B. [3]

   iii) State with a reason whether events A and B are mutually exclusive. [2]

   Cambridge International AS and A Level Mathematics 9709, Paper 61 Q4 October/November 2013

14. Boxes of sweets contain toffees and chocolates. Box A contains 6 toffees and 4 chocolates, box B contains 5 toffees and 3 chocolates, and box C contains 3 toffes and 7 chocolates. One of the boxes is chosen at random and two sweets are taken out, one after the other, and eaten.

   i) Find the probability that they are both toffees. [3]

   ii) Given that they are both toffees, find the probability that they both came from box A. [3]

   Cambridge International AS and A Level, Paper 6 Q2 October/November 2005

15. A choir consists of 13 sopranos, 12 altos, 6 tenors and 7 basses. A group consisting of 10 sopranos, 9 altos, 4 tenors and 4 basses is to be chosen from the choir.

   i) In how many different ways can the group be chosen? [2]

   ii) In how many ways can the 10 chosen sopranos be arranged in a line if the 6 tallest stand next to each other? [3]

   iii) The 4 tenors and 4 basses in the group stand in a single line with all the tenors next to each other and all the basses next to each other. How many possible arrangements are there if three of the tenors refuse to stand next to any of the basses? [3]

Cambridge International AS and A Level Mathematics 9709, Paper 6 Q4 May/June 2009

16. At a zoo, rides are offered on elephants, camels and jungle tractors. Ravi has money for only one ride. To decide which ride to choose, he tosses a fair coin twice. If he gets 2 heads he will go on the elephant ride, if he gets 2 tails he will go on the camel ride and if he gets 1 of each he will go on the jungle tractor ride.

   i) Find the probabilities that he goes on each of the three rides. [2]

The probabilities that Ravi is frightened on each of the three rides are as follows:

Elephant ride $\frac{6}{10}$, camel ride $\frac{7}{10}$, jungle tractor ride $\frac{8}{10}$.

   ii) Draw a fully labelled tree diagram showing the rides that Ravi could take and whether or not he is frightened. [2]

Ravi goes on a ride.

   iii) Find the probability that he is frightened. [2]

   iv) Given that Ravi is **not** frightened, find the probability that he went on the camel ride. [3]

Cambridge International AS and A Level Mathematics 9709, Paper 6 Q5 May/June 2009

17. When Don plays tennis, 65% of his first serves go into the correct area of the court. If the first serve goes into the correct area, his chance of winning the point is 90%. If his first serve does not go into the correct area, Don is allowed a second serve, and of these, 80% go into the correct area. If the second serve goes into the correct area, his chance of winning the point is 60%. If neither serve goes into the correct area, Don loses the point.

   i) Draw a tree diagram to represent this information. [4]

   ii) Using your tree diagram, find the probability that Don loses the point. [3]

   iii) Find the conditional probability that Don's first serve went into the correct area, given that he loses the point. [2]

Cambridge International AS and A Level Mathematics 9709, Paper 6 Q6 May/June 2004

18. Box *A* contains 5 red paper clips and 1 white paper clip. Box *B* contains 7 red paper clips and 2 white paper clips. One paper clip is taken at random from box *A* and transferred to box *B*. One paper clip is then taken at random from box *B*.

   i) Find the probability of taking both a white paper clip from box *A* and a red paper clip from box *B*. [2]

   ii) Find the probability that the paper clip taken from box *B* was red. [2]

   iii) Find the probability that the paper clip taken from box *A* was red, given that the paper clip taken from box *B* is red. [2]

   iv) The random variable *X* denotes the number of times that a red paper clip is taken. Draw up a table to show the probability distribution of *X*. [4]

Cambridge International AS and A Level Mathematics 9709, Paper 6 Q7 October/November 2007

19. A vegetable basket contains 12 pepers, of which 3 are red, 4 are green and 5 are yellow. Three peppers are taken, at random and without replacement, from the basket.

i) Find the probability that the three peppers are all different colours. [3]

ii) Show that the probability that exactly 2 of the peppers taken are green is $\frac{12}{55}$.

iii) The number of green peppers taken is denoted by the discrete random variable X. Draw up a probability distribution table for X. [5]

Cambridge International AS and A Level Mathematics 9709, Paper 61 Q7 May/June 2006

20. The discrete random variable X has the following probability distribution.

x	0	1	2	3	4
$P(X = x)$	0.26	q	$3q$	0.05	0.09

i) Find the value of q. [2]

ii) Find $E(X)$ and $Var(X)$ [3]

Cambridge International AS and A Level Mathematics 9709, Paper 6 Q2 October/ November 2006

21. A box contains five balls numbered 1, 2, 3, 4, 5. Three balls are drawn randomly at the same time from the box.

i) By listing all possible outcomes (123, 124, etc.), find the probability that the sum of the three numberes drawn is an odd numbers. [2]

The random variable L denotes the largest of the three numbers drawn.

ii) Find the probability that L is 4. [1]

iii) Draw up a table to show the probability distribution of L. [3]

iv) Calculate the expectation and variance of L. [3]

Cambridge International AS and A Level Mathematics 9709, Paper 6 Q6 October/ November 2004

22. Two fair dice are thrown. Let the random variable X be the smaller of the two scores if the scores are different, or the score on one of the dice if the scores are the same.

i) Copy and complete the following table to show the probability distribution of X. [3]

x	1	2	3	4	5	6
$P(X = x)$						

ii) Find $E(X)$ [2]

Cambridge International AS and A Level Mathematics 9709, Paper 6 Q3 May/June 2004

23. Every day Eduardo tries to phone his friend. Every time he phones there is a 50% chance that his friend will answer. If his friend answers, Eduardo does not phone again on that day. If his friend does not answer, Eduardo tries again in a few minutes' time. If his friend has not answered after 4 attempts, Eduardo does not try again on that day.

i) Draw a tree diagram to illustrate this situation. [3]

ii) Let X be the number of unanswered phone calls made by Eduardo on a day. Copy and complete the table showing the probability distribution of X. [4]

x	0	1	2	3	4
$P(X = x)$		$\frac{1}{4}$			

Cambridge International AS and A Level Mathematics 9709, Paper 6 Q6 i and ii May/June 2008

24. Two unbiased tetrahedral dice each have four faces numbered 1, 2, 3 and 4. The two dice are thrown together and the sum of the numbers on the faces on which they land is noted. Find the expected number of occasions on which this sum is 7 or more when the dice are thrown together 200 times.

Cambridge International AS and A Level Mathematics 9709, Paper 62 Q2 October/ November 2009

25. The possible values of the random variable X are the 8 integers in the set $\{-2, -1, 0, 1, 2, 3, 4, 5\}$. The probability of X being 0 is $\frac{1}{10}$. The probabilities for all the other values of X are equal. Calculate

i) $P(X < 2)$, [2]

ii) the variance of X, [3]

iii) the value of a for which
$P(-a \le X \le 2a) = \frac{17}{35}$. [1]

Cambridge International AS and A Level Mathematics 9709, Paper 61 Q3 May/June 2011

26. A box contains 10 pens of which 3 are new. A random sample of two pens is taken.

i) Show that the probability of getting exactly one new pen in the sample is $\frac{7}{15}$. [2]

(ii) Construct a probability distribution table for the number of new pens in the sample. [3]

iii) Calculate the expected number of new pens in the sample. [1]

Cambridge International AS and A Level Mathematics 9709, Paper 6 Q2 May/June 2003

27. In a particular discrete probability distribution the ramdom variable X takes the value $\frac{120}{r}$ with probability $\frac{r}{45}$, where r takes all integer values from 1 to 9 inclusive.

i) Show that $P(X = 40) = \frac{1}{15}$. [2]

ii) Construct the probability distribution table for X. [3]

iii) Which is the modal value of X? [1]

iv) Find the probability that X lies between 18 and 100. [2]

Cambridge International AS and A Level Mathematics 9709, Paper 62 Q5 October/ November 2009

28. Six men and three women are standing in a supermarket queue.

i) How many possible arrangements are there if there are no restrictions on order? [2]

ii) How many possible arrangements are there if no two of the women are standing next to each other? [4]

iii) Three of the people in the queue are chosen to take part in a customer survey. How many different choices are possible if at least one woman must be included? [3]

Cambridge International AS and A Level Mathematics 9709, Paper 6 Q6 October/ November 2006

29. a) i) Find how many different four-digit numbers can be made using only the digits 1, 3, 5 and 6 with no digit being repeated. [1]

ii) Find how many different odd numbers greater than 500 can be made using some or all of the digits 1, 3, 5 and 6 with no digit being repeated. [4]

b) Six cards numbered 1, 2, 3, 4, 5, 6 are arranged randomly in a line. Find the probability that the cards numbered 4 and 5 are **not** next to each other. [3]

Cambridge International AS and A Level Mathematics 9709, Paper 62 Q4 October/ November 2009

30. ● ● ● ●

Pegs are to be placed in the four holes shown, one in each hole. The pegs come in different colours and pegs of the same colour are identical. Calculate how many different arrangements of coloured pegs in the four holes can be made using

i) 6 pegs, all of different colours,

ii) 4 pegs consiting of 2 blue pegs, 1 orange peg and 1 yellow peg. [1]

Beryl has 12 pegs consisting of 2 red, 2 blue, 2 green, 2 orange, 2 yellow and 2 black pegs. Calculate how many different arrangements of coloured pegs in the 4 holes Beryl can make using

iii) 4 different colours, [1]

iv) 3 different colours, [3]

v) any of her 12 pegs. [3]

Cambridge International AS and A Level Mathematics 9709, Paper 61 Q6 October/ November 2010

31. a) Find the number of different ways in which the 12 letters of the word STRAWBERRIES can be arranged

    i) if there are no restrictions, [2]

    ii) if the 4 vowels, A, E, E, I must all be together. [3]

b) **i)** 4 astronauts are chosen from a certain number of candidates. If order of choosing is not taken into account, the number of ways the astronauts can be chosen is 3876. How many ways are there if order of choosing is taken into account? [2]

    ii) 4 astronauts are chosen to go on a mission. Each of these astronauts can take 3 personal possessions with him. How many different ways can these 12 possessions be arranged in row if each astronaut's possessions are kept together? [2]

Cambridge International AS and A Level Mathematics 9709, Paper 61 Q6 October/ November 2011

32. A builder is planning to build 12 houses along one side of a road. He will build 2 houses in style A, 2 houses in style B, 3 houses in style C, 4 houses in style D and 1 house in style E.

i) Find the number of possible arrangements of these 12 houses. [2]

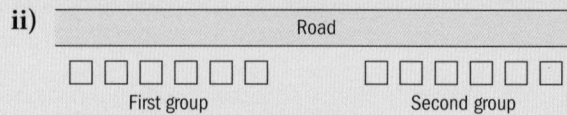

ii)

Road

☐☐☐☐☐☐   ☐☐☐☐☐☐
First group   Second group

The 12 houses will be in two groups of 6 (see diagram). Find the number of possible arrangements if all the houses in styles A and D are in the first group and all the houses in styles B, C and E are in the second group. [3]

iii) Four of the 12 houses will be selected for a survey. Exactly one house must be in style B and exactly one house in style C. Find the number of ways in which these four houses can be selected. [2]

Cambridge International AS and A Level Mathematics 9709, Paper 6 Q4 October/November 2008

33. The discrete random variable X has the following probability distribution.

x	1	3	5	7
$P(X = x)$	0.3	a	b	0.25

i) Write down an equation satisfied by a and b. [1]

ii) Given that $E(X) = 4$, find a and b. [3]

Cambridge International AS and A Level Mathematics 9709, Paper 6 Q1 October/November 2002

34. Ivan throws three fair dice.

   i) List all the possible scores on the three dice which given a total score of 5, and hence show that the probability of Ivan obtaining a total score of 5 is $\frac{1}{36}$.   [3]

   ii) Find the probability of Ivan obtaining a total score of 7.   [3]

Cambridge International AS and A Level Mathematics 9709, Paper 6 Q2 October/ November 2002

35. In a certain hotel, the lock on the door to each room can be opened by inserting a key card. The key card can be inserted only one way round. The card has a pattern of holes punched in it. The card has 4 columns, and each column can have either 1 hole, 2 holes, 3 holes or 4 holes punched in it. Each column has 8 different positions for the holes. The diagram illustrates one particular key card with 3 holes punched in the first column, 3 in the second, 1 in the third and 2 in the fourth.

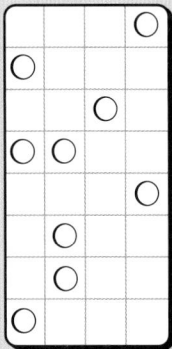

   i) Show that the number of different ways in which a column could have exactly 2 holes is 28.   [1]

   ii) Find how many different patterns of holes can be punched in a column.   [4]

   iii) How many different possible key cards are there?   [2]

Cambridge International AS and A Level Mathematics 9709, Paper 6 Q4 October/ November 2002

36. a) A collection of 18 books contains one Harry Potter book. Linda in going to choose 6 of these books to take on holiday.

   i) In how many ways can she choose 6 books?   [1]

   ii) How many of these choices will include the Harry Potter book?   [2]

  b) In how many ways can 5 boys and 3 girls stand in a straight line

   i) if there are no restrictions,   [1]

   ii) if the boys stand next to each other?   [4]

Cambridge International AS and A Level Mathematics 9709, Paper 6 Q6 October/ November 2003

37. A discrete random variable X has the following probability distribution.

x	1	2	3	4
$P(X = x)$	$3c$	$4c$	$5c$	$6c$

   i) Find the value of the constant c.   [2]

   ii) Find $E(X)$ and $Var(X)$.   [4]

   iii) Find $P(X > E(X))$ [2]

Cambridge International AS and A Level Mathematics 9709, Paper 6 Q8 October/ November 2003

38. Four families go the a theme park together. Mr and Mrs Lin take their 2 children. Mr O'Connor takes his 2 children. Mr and Mrs Ahmed take their 3 children. Mrs Burton takes her son. They 14 people all have to go through a turnstile one at a time to enter the theme park.

   i) In how many different orders can the 14 people go through the turnstile if each family stays together?   [3]

   ii) In how many diffferent orders can the 8 children and 6 adults go through the turnstile if no two adults go consecutively? One inside the theme park, the children go on the roller-coaster. Each roller-coaster car holds 3 people.

iii) In how many different ways can the 8 children be divided into two groups of 3 and one group of 2 to go on the roller-coaster?

Cambridge International AS and A Level Mathematics 9709, Paper 61 Q6 May/June 2013

39. The people living in two towns, Mumbok and Bagville, are classified by age. The numbers in thousands living in each town are shown in the table below.

	Mumbok	Bagville
Under 18 years	15	35
18 to 60 years	55	95
Over 60 years	20	30

One of the towns is chosen. The probability of choosing Mumbok is 0.6 and the probability of choosing Bagville is 0.4. Then a person is chosen at random from that town. Given that the person chosen is between 18 and 60 years old, find the probability that the town chosen was Mumbok. [5]

Cambridge International AS and A Level Mathematics 9709, Paper 61 Q2 October/November 2013

40. A shop has 7 different mountain bicycles, 5 different racing bicycles and 8 different ordinary bicycles on display. A cycling club selects 6 of these 20 bicycles to buy.

i) How many different selections can be made if there must be no more than 3 mountain bicycles and no more than 2 of each of the other types of bicycle? [4]

The cycling club buys 3 mountain bicycles, 1 racing bicycle and 2 ordinary bicycles and parks them in a cycle rack, which has a row of 10 empty spaces.

ii) How many different arrangements are there in the cycle rack if the mountain bicycles are all together with no spaces between them, the ordinary bicycles are both together with no spaces between them and the spaces are all together? [3]

iii) How many different arrangements are there in the cycle rack if the ordinary bicycles are at each end of the bicycles and there are no spaces between any of the bicucles? [3]

Cambridge International AS and A Level Mathematics 9709, Paper 61 Q6 October/November 2013

41. James has a fair coin and a fair tetrahedral die with four faces numbered 1, 2, 3, 4. He tosses the coin once and the die twice. The random variable X is defined as follows.

- If the coin shows a **head** then X is the **sum** of the scores on the two throws of the die.
- If the coin shows a **tail** then X is the score on the **first throw** of the die only.

i) Explain why $X = 1$ can only be obtained by throwing a tail, and show that $P(X = 1) = \frac{1}{8}$. [2]

ii) Show that $P(X = 3) = \frac{3}{16}$. [4]

iii) Copy and complete the probability distribution table for X. [3]

x	1	2	3	4	5	6	7	8
$P(X = x)$	$\frac{1}{8}$		$\frac{3}{16}$		$\frac{1}{8}$		$\frac{1}{16}$	$\frac{1}{32}$

Event Q is 'James throws a tail'. Event R is 'the value of X is 7'.

iv) Determine whether events Q and R are exclusive. Justify your answer. [2]

Cambridge International AS and A Level Mathematics 9709, Paper 61 Q7 October/November 2013

42. The probability that Henk goes swimming on any day is 0.2. On a day when he goes swimming, the probability that Henk has burgers for supper is 0.75. On a day when he does not go swimming the probability that he has burgers for supper is x. This information is shown on the following tree diagram.

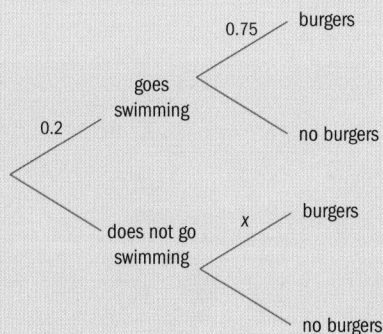

The probability that Henk has burgers for supper on any day is 0.5.

i) Find x.

ii) Given that Henk has burgers for supper, find the probability that he went swimming that day.

Cambridge International AS and A Level Mathematics 9709, Paper 6 Q2 May/June 2006

43. When Andrea needs a taxi, she rings one of three taxi companies, A, B or C. 50% of her calls are to taxi company A, 30% to B and 20% to C. A taxi from company A arrives late 4% of the time, a taxi from company B arrives late 6% of the time and a taxi from company C arrives late 17% of the time.

i) Find the probability that, when Andrea rings for a taxi, it arrives late.

ii) Given that Andrea's taxi arrives late, find the conditional probability that she rang company B.

Cambridge International AS and A Level Mathematics 9709, Paper 6 Q3 October/November 2004

Maths in real-life

Sporting statistics

Scoring systems are used in all sports. In multi-disciplinary events, such as the decathlon and the heptathlon, these systems are particularly complex and make use of statistical formulae.

Both the decathlon and heptathlon have to transform performance in time for running different events, and distances in throwing or jumping events into a point scored for each component event. The winner is the person scoring the highest total number of points in the events.

Ashton Eaton set a world record in the decathlon in June 2012, scoring 9,039 points in ten events over two days.

Jessica Ennis ran the fastest ever 100 metres hurdles in the 2012 Olympics on her way to winning the gold medal with a score of 6995 points from the 7 events.

Some heptathletes and decathletes are stronger in sprints, some in distance running and others in throwing events, and the scoring system has to try to reward performances in all disciplines on a reasonable basis.

In running, good performances are low times while in throwing and jumping the good performances are larger distances or heights, so a different approach is needed for these two groups of events and different formulae are required for each.

This table shows the points earned in each discipline by the 5 best decathletes of all time (up to July 2014). You can see that there is considerable variation between athletes in a discipline and there is also variation between each discipline.

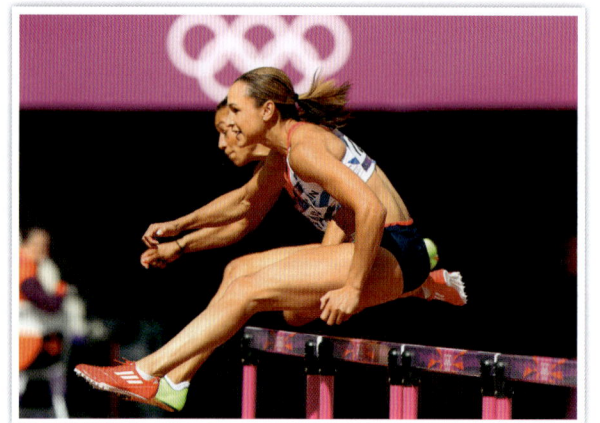

The full data set can be found at the end of the book.

Athlete	100m	Long	Shot	High	400m	110h	Discus	Pole	Javelin	1500m	Total
Ashton Eaton	1044	1120	741	850	973	1014	722	1004	721	850	9039
Roman Šebrle	942	1089	810	915	919	985	827	849	892	798	9026
Tomaš Dvorak	966	1035	899	840	905	1010	836	880	925	698	8994
Dan O'Brien	992	1081	894	868	885	977	840	910	777	667	8891
Daley Thompson	989	1063	834	831	960	932	799	910	817	712	8847

This graph shows the average score in each of the ten decathlon events for the top 75 personal best scores (i.e. these scores are from 75 different athletes). You can see that there is a considerable amount of variation across events.

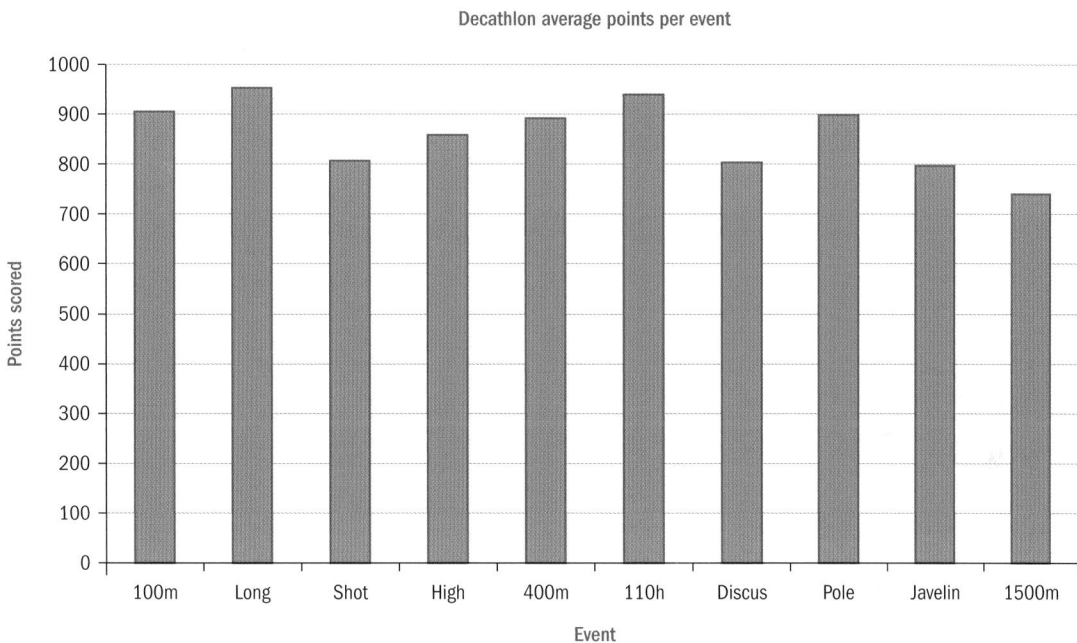

Decathlon average points per event

Is this fair?

The formulae used to calculate points for each discipline have remained the same since 1984. In that time there have been enormous increases in sports and medical sciences as well as in coaching regimes, and the advent of professional athletes. Some events are very technical, while others rely on differing combinations of strength, speed and stamina. The widely differing average scores taken over the 75 best ever decathletes suggests that performances in some events are rewarded differently, but changing the scoring system is not an easy task when these are the yardsticks against which all decathletes train. A similar story holds for the women's heptathlon.

The binomial distribution

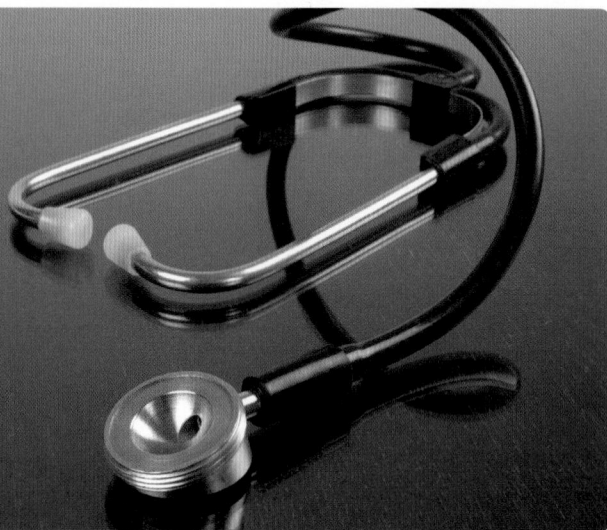

The binomial distribution can be used to analyze any situation where outcomes of interest can be classified into two categories – either something happens, or it doesn't. The observed number of times something has happened can then be compared with a model of what is expected to happen under normal circumstances. If what has been seen is unusual, for example a doctor has an unusually high number of patients dying or having to be transferred to hospital, then further investigation can take place. The explanation may be that the doctor is incompetent, or it even may lie in a doctor's excellence.

Objectives

- Use formulae for probabilities for the binomial distribution.
- Recognise practical situations where the binomial distribution is a suitable model.
- Use the notation $X \sim B(n, p)$.
- Use formulae for the mean and variance of the binomial distribution.

Before you start

You should know how to:

1. Use your calculator to work out factorials and values of $\binom{n}{r}$,

   e.g. find the value of:

   a) $6!$ b) $\binom{8}{3}$

   a) $6! = 6 \times 5 \times 4 \times 3 \times 2 \times 1 = 720$

   b) $\binom{8}{3} = \dfrac{8!}{3!5!} = \dfrac{(8 \times 7 \times 6 \times 5 \times 4 \times 3 \times 2 \times 1)}{(3 \times 2 \times 1)(5 \times 4 \times 3 \times 2 \times 1)} = 56$

2. Substitute values into simple formulae,

   e.g. find the value of $V = np(1 - p)$ when $n = 20$ and $p = 0.3$
   $V = 20 \times 0.3(1 - 0.3) = 4.2$

Skills check:

1. Find the value of

   a) $10!$ b) $\binom{15}{7}$.

2. Find the value of $M = kx(3 - x)$ when $k = 10$ and $x = 0.4$

7.1 Introducing the binomial distribution

You can use tree diagrams to work out probabilities of events involving a number of separate stages. A special case is where each stage has only two outcomes of interest, and the probabilities of these two outcomes are the same at each stage.

Even when there are more than two basic outcomes, they can usually be grouped into two categories: those which satisfy the event of interest (often called 'success'), and those which do not ('failure'). At any stage, $P(\text{failure}) = 1 - P(\text{success})$.

For example, if the aim was to count how often a factor of 6 is thrown when a die is rolled, then throws of any of 1, 2, 3 and 6 will result in 'success'.

So, in this case: $P(\text{success}) = \frac{2}{3}$ and $P(\text{failure}) = \frac{1}{3}$.

Some patterns begin to emerge when these probabilities are shown on the branches of a tree diagram and the number of successes is counted:

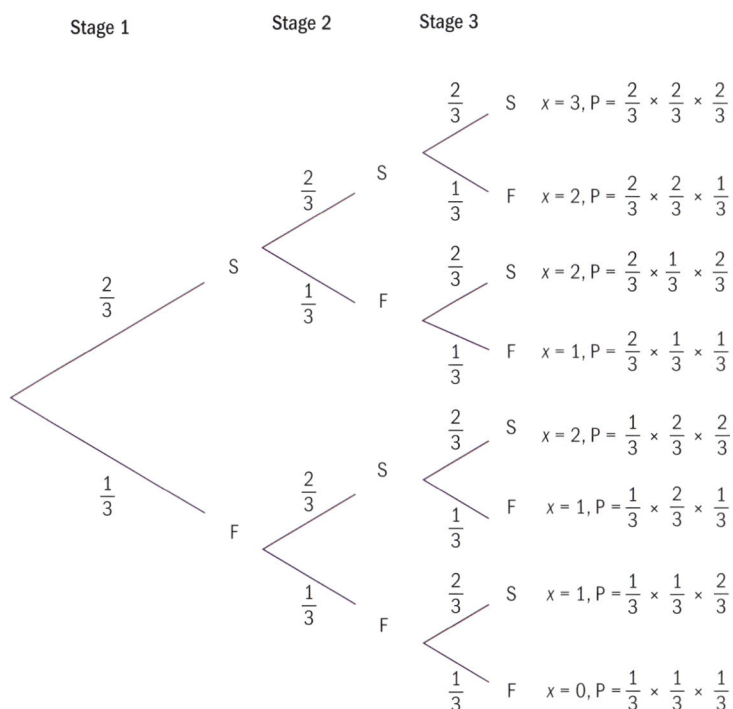

Stage 1   Stage 2   Stage 3

$$S \quad x = 3, P = \frac{2}{3} \times \frac{2}{3} \times \frac{2}{3}$$
$$F \quad x = 2, P = \frac{2}{3} \times \frac{2}{3} \times \frac{1}{3}$$
$$S \quad x = 2, P = \frac{2}{3} \times \frac{1}{3} \times \frac{2}{3}$$
$$F \quad x = 1, P = \frac{2}{3} \times \frac{1}{3} \times \frac{1}{3}$$
$$S \quad x = 2, P = \frac{1}{3} \times \frac{2}{3} \times \frac{2}{3}$$
$$F \quad x = 1, P = \frac{1}{3} \times \frac{2}{3} \times \frac{1}{3}$$
$$S \quad x = 1, P = \frac{1}{3} \times \frac{1}{3} \times \frac{2}{3}$$
$$F \quad x = 0, P = \frac{1}{3} \times \frac{1}{3} \times \frac{1}{3}$$

Consider the paths with $x = 2$: there are three of them, and they all have $\frac{2}{3}, \frac{2}{3}$ and $\frac{1}{3}$ multiplied together, but in different orders.

All three paths with $x = 1$ have $\frac{1}{3}$, $\frac{1}{3}$ and $\frac{2}{3}$ multiplied together, but again in different orders.

There is only one path for each of $x = 0$ and $x = 3$. Respectively, they have

$\frac{1}{3}$, $\frac{1}{3}$ and $\frac{1}{3}$, and $\frac{2}{3}$, $\frac{2}{3}$ and $\frac{2}{3}$ multiplied together.

The number of paths corresponds to the number of different ways you can order the three probabilities, in each case. The 1, 3, 3, 1 sequence of the number of paths, for $x = 0$, 1, 2, 3, is one of the rows of Pascal's triangle, as shown on the right.

```
            1
          1   1
        1   2   1
      1   3   3   1
    1   4   6   4   1
  1   5  10  10   5   1
1   6  15  20  15   6   1
```

You can write the probabilities for the number of successes in three throws of the die like this:

$$P(x = 0) = 1 \times \left(\frac{1}{3}\right)^3$$

$$P(x = 1) = 3 \times \frac{2}{3} \times \left(\frac{1}{3}\right)^2$$

$$P(x = 2) = 3 \times \left(\frac{2}{3}\right)^2 \times \frac{1}{3}$$

$$P(x = 3) = 1 \times \left(\frac{2}{3}\right)^3$$

If the tree diagram is extended to six stages (that is, six throws of the die), there would be 64 different paths, and it would be cumbersome to work through the whole process like this.

However, the coefficients in Pascal's triangle and the patterns seen in the diagram with three stages can be used to write down expressions for the probabilities of getting 0, 1, 2, 3, 4, 5 or 6 successes:

x	0	1	2	3	4	5	6
$P(x = x)$	$\left(\frac{1}{3}\right)^6$	$6\left(\frac{1}{3}\right)^5\left(\frac{2}{3}\right)$	$15\left(\frac{1}{3}\right)^4\left(\frac{2}{3}\right)^2$	$20\left(\frac{1}{3}\right)^3\left(\frac{2}{3}\right)^3$	$15\left(\frac{1}{3}\right)^2\left(\frac{2}{3}\right)^4$	$6\left(\frac{1}{3}\right)\left(\frac{2}{3}\right)^5$	$\left(\frac{2}{3}\right)^6$

Example 1

If the probability that Saanvi is late home from work on any day is 0.4, what is the probability she is late home twice in a 5-day working week?

· ·

The relevant row in Pascal's triangle is 1, 5, 10, 10, 5, 1, where the numbers represent how many ways you can get 0, 1, 2, 3, 4 and 5 out of 5. There are 10 ways of getting 2, so

P(late twice) = $10 \times 0.4^2 \times 0.6^3$.

> If Saanvi is late twice in the week she will not be late the other three times – this gives the powers of 0.4 (late) and 0.6 (not late).

Example 2

Adisa tries to eat at least five pieces of fruit each day. The probability that he does is 0.6, independent of any other day. What is the probability that Adisa eats at least five pieces of fruit on more than half the days in a given week.

· ·

The relevant row in Pascal's triangle is the next one:

$$1, 7, 21, 35, 35, 21, 7, 1$$

P(more than half the days)

$= 35 \times 0.6^4 \times 0.4^3 + 21 \times 0.6^5 \times 0.4^2 + 7 \times 0.6^6 \times 0.4 + 0.6^7$

> More than half the days in a week means at least 4 days.

When the number of stages (or trials) is large (> 6), it is hard to use Pascal's triangle to find the binomial coefficients.

> You met this in Chapter 6. Remember that the value of 0! is defined as 1

The number of paths giving r occurrences out of n cases equals the number of ways of choosing r out of n, which is

$$^nC_r = \binom{n}{r} = \frac{n!}{r!(n-r)!},$$

and can also be found in the nth row of Pascal's triangle (the row starting 1, n, …).

> Most scientific calculators have a button marked ^nC_r

The binomial probability distribution is defined as

$$P(X=r) = \binom{n}{r} p^r q^{n-r} \quad r = 0, 1, 2, \ldots, n \quad q = 1 - p.$$

We need to have values for n, the number of trials, and for p, the probability of 'success' on any one trial, in order for this to make sense, so there is a family of binomial distributions with two parameters.

The parameters of the binomial distribution are n, the number of trials, and p, the probability of a 'success' on any one trial.

For simplicity, the distribution is often written as $X \sim B(n, p)$.

For Example 1, if X is the number of times Saanvi is late home in a week, then $X \sim B(5, 0.4)$

Example 3

If $X \sim B(12, 0.2)$, find the probability that $X = 3$.

$$P(X=3) = \binom{12}{3}(0.2)^3 (0.8)^9 = 0.236 \text{ (to 3 s.f.)}$$

Example 4

If $X \sim B(10, 0.9)$ find the probability that $X \leq 8$.

Rather than calculate all of the probabilities that X is 0, 1, 2, …, 8, we can calculate $P(X = 9 \text{ or } 10)$ as the complementary probability and subtract this from 1.

$$P(X \leq 8) = 1 - \left[\binom{10}{9}(0.9)^9 (0.1) + (0.9)^{10}\right] = 0.264 \text{ (to 3 s.f.)}$$

Example 5

If 25 dice are rolled, find the probability that three 6s are seen.

If X = number of 6s seen, then $X \sim B\left(25, \frac{1}{6}\right)$.

$$P(X=3) = \binom{25}{3}\left(\frac{1}{6}\right)^3 \left(\frac{5}{6}\right)^{22} = 0.193 \text{ (to 3 s.f.)}$$

It is worth practising using the formula so you are confident and accurate when you need to use it.

You can use a spreadsheet or handheld calculator to work out binomial probabilities by specifying n, p and the value of X.

Exercise 7.1

1. **a)** The first few rows of Pascal's triangle are shown here.
   Write down the next row.

   ```
                1
              1   1
            1   2   1
          1   3   3   1
        1   4   6   4   1
      1   5  10  10   5   1
    1   6  15  20  15   6   1
   ```

   b) Work out the values of

   i) $\binom{7}{3}$ **ii)** $^7C_5$

   c) Make sure you know which elements in Pascal's triangle in part **(a)** correspond to the calculations in part **(b)**.

2. **a)** Calculate

   i) $\binom{10}{4}$ **ii)** $\binom{9}{0}$

   iii) $\binom{15}{6}$ **iv)** $\binom{100}{2}$

   b) Calculate

   i) $^{10}C_2$ **ii)** $^{11}C_6$
   iii) $^{12}C_{12}$ **iv)** $^{50}C_{20}$

3. If $X \sim B(6, 0.5)$, find the probability that
   a) $X = 2$ **b)** $X = 5$.

4. If $X \sim B(6, 0.3)$, find the probability that
   a) $X = 2$ **b)** $X = 5$.

5. If $X \sim B(12, 0.4)$, find the probability that
   a) $X = 3$ **b)** $X = 8$.

6. If $X \sim B(12, 0.7)$, find the probability that
   a) $X = 3$ **b)** $X = 8$.

7. If $X \sim B\left(10, \frac{1}{3}\right)$, find the probability that

   a) $X = 0$ **b)** $X = 5$.

8. If $X \sim B\left(15, \frac{3}{4}\right)$, find the probability that

   a) $X = 7$ **b)** $X = 8$.

9. If a fair coin is tossed six times, calculate the probability of seeing
   a) 2 heads **b)** 5 heads.

10. If a fair coin is tossed ten times, calculate the probability of seeing

    a) **i)** 0 heads   **ii)** 1 head

        iii) 2 heads   **iv)** 3 heads.

    b) Using your answers to part (**a**), calculate the probability of seeing

        i)   no more than 1 head

        ii) at least 3 heads

        iii) more than 3 heads.

11. If $X \sim B(12, 0.15)$, find

    a) $P(X > 2)$ **b)** $P(X \leq 10)$.

12. If $X \sim B\left(20, \dfrac{1}{4}\right)$ find

    a) $P(X \leq 2)$ **b)** $P(X > 19)$.

13. Suki was absent when the class were told they would have a test during the next lesson, so she has done no revision and has to guess at the answers to five multiple-choice questions. If there are four choices for each question, find the probability she gets

    a) none right

    b) three right.

7.2 Mean and variance of the binomial distribution

How many 'fives' would you expect to get, on average, if you rolled a fair die 120 times?

With a fair die, the probability of getting a five on any go is $\dfrac{1}{6}$, so 'on average' you would expect to see $\dfrac{1}{6}$ of the rolls come up with a five; that is, on average you would expect 20 fives in 120 throws.

> This does not imply that you expect to get exactly 20 fives if you roll a fair die 120 times

If you did a large number of sets of n trials of a binomial probability distribution, $X \sim B(n, p)$, then the average number of successes out of n trials would allow you to estimate the value of p. If p is the probability that any individual trial gives a success then the proportion of successes in the long term will be equal to p.

Did you know?
One important property of random processes is that long-term behaviour is quite predictable, although short-term behaviour is consistently unpredictable. The 'law of large numbers' sets out these properties, but is beyond the scope of this book.

If $X \sim B(n, p)$, then mean, $E(X) = np$, and variance, $Var(X) = (\sigma^2) = npq$, where $q = 1 - p$.

You must be able to use these results but are not required to prove them formally.

Example 6

If $X \sim B(10, 0.2)$ find the mean and variance of X.

$n = 10$ and $p = 0.2$, so $q = 0.8$,

and $np = 2$, $npq = 1.6$.

Therefore the mean is 2 and the variance is 1.6.

Example 7

If $X \sim B(80, 0.4)$ find the mean and standard deviation of X.

$n = 80$ and $p = 0.4$, so $q = 0.6$,

and $np = 32$, $npq = 19.2$.

Therefore the mean is 32 and the variance is 19.2, so the standard deviation is $\sqrt{19.2}$.

Example 8

X is a binomial distribution with mean 8 and variance 6.4. Find $P(X \leq 1)$.

$np = 8$, $npq = 6.4 \Rightarrow q = 0.8 \Rightarrow p = 0.2$, $n = 40$.

Using the $X \sim B(40, 0.2)$ distribution we get $P(X \leq 1) = (0.8)^{40} + 40(0.2)(0.8)^{39} = 0.00146$ (to 3 s.f.)

Exercise 7.2

1. If $X \sim B(35, 0.5)$, find
   a) $E(X)$ b) $Var(X)$

2. If $X \sim B(75, 0.4)$, find the mean and standard deviation of X.

3. X is a binomial random variable based on 25 trials. $E(X) = 20$.
   a) Find the standard deviation of X.
   b) Find $P(X > \mu)$, where $\mu = E(X)$.

4. X is a binomial random variable with $E(X) = 30$ and $Var(X) = 21$.
Find the number of trials.

5. Jakob is a dentist who finds that, on average, one-in-five of his patients does not turn up for their appointment.

  a) Using a binomial distribution, find the mean and variance of the number of patients who turn up for their appointments, in a clinic with 20 appointments.

  b) Find the probability that more than three patients do not turn up for their appointment in that clinic.

6. If $X \sim B(20, p)$ find $Var(X)$ in terms of p. Find the value of p which gives the largest variance.

7. If $X \sim B(8, 0.12)$ find $P(X < \mu - \sigma)$ where $\mu = E(X)$, $\sigma = \sqrt{Var(X)}$.

7.3 Modelling with the binomial distribution

If a coin is tossed 20 times, how likely would it be to get exactly 10 heads?

If someone argued that there is less than a 50% chance that a fair coin will show 9, 10 or 11 heads when it is tossed 20 times, how accurate would they be?

The formula will show that the probability is actually 0.4966

The graphs that follow show the probabilities for a number of binomial distributions. They show the way probabilities in the binomial family behave.

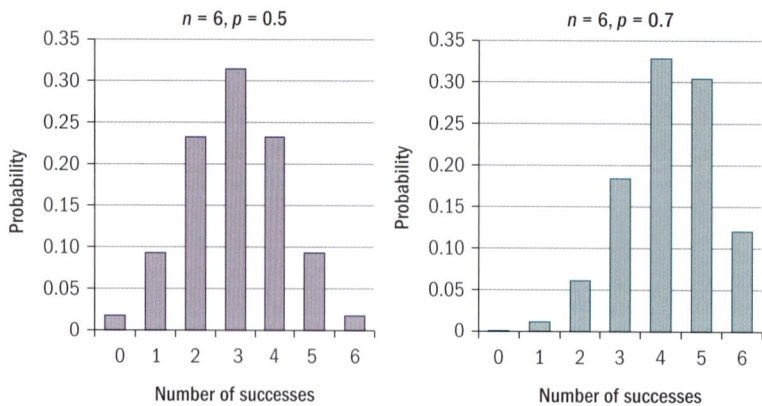

When $p = 0.5$ the distribution is symmetrical and peaks at three successes out of six trials.

When $p = 0.7$ successes would be expected to come more often and the distribution peaks at four successes out of six trials.

There are about three or four of the probabilities in these cases that would be described as relatively high. The largest individual probabilities can be estimated as being around $\frac{1}{3}$.

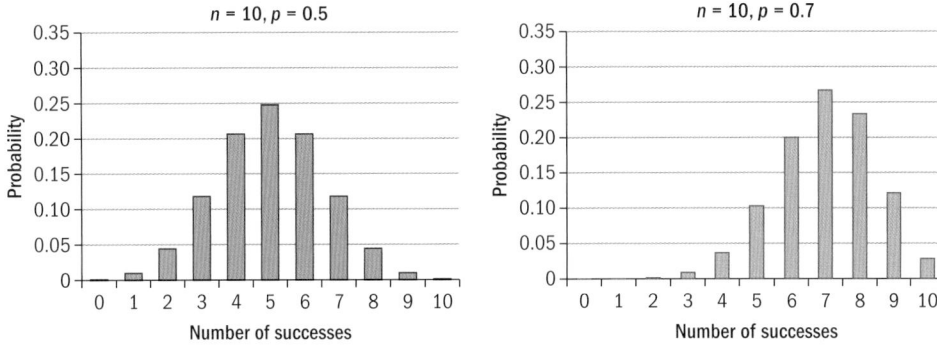

n = 10, p = 0.5

n = 10, p = 0.7

If n is increased to 10, there are a larger number of the possible outcomes which have relatively high probabilities, but since the total probability is 1 it follows that the largest probabilities cannot be as high as in the previous cases. With p slightly larger than 0.5, the distribution will be slightly skewed, with a tail of smaller probabilities for low numbers of successes.

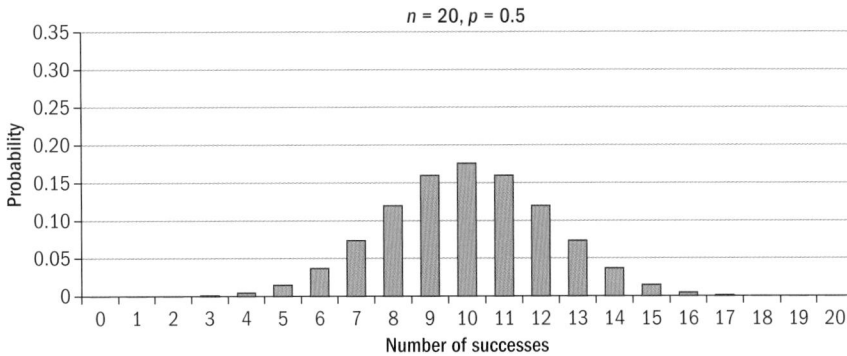

n = 20, p = 0.5

With a larger value of n (here 20) and $p = 0.5$, the graph is symmetric, with again about half of the possible number of successes contributing relatively high probabilities. The highest single probability is less than $\frac{1}{5}$, and the total probability of seeing 9, 10 or 11 successes is less than $\frac{1}{2}$.

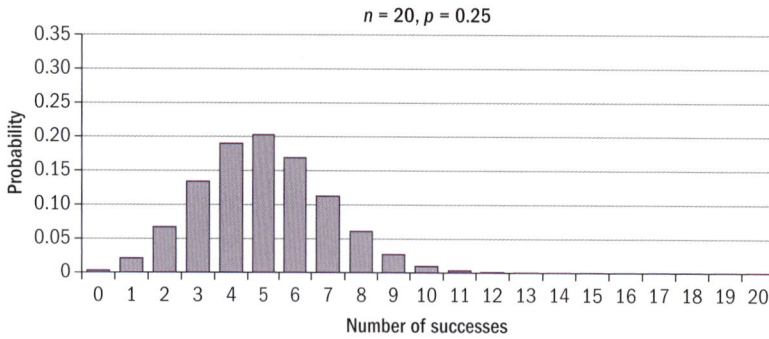

n = 20, p = 0.25

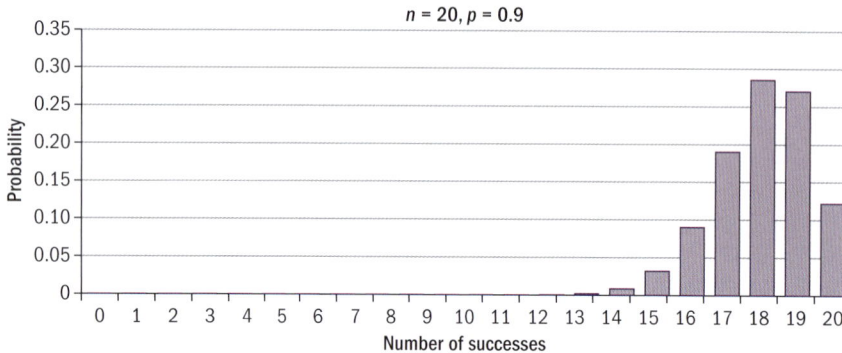

n = 20, p = 0.9

When the value of p is moved down or up, the peak of the distribution is moved left or right, respectively. When p moves further from $\frac{1}{2}$, the distribution peak gets higher and the distribution becomes less spread out (there is lower variance).

The binomial distribution can be used for any situation where the aim is to count the number of times a particular outcome is observed out of a fixed number of cases – provided certain conditions are satisfied.

The conditions for the binomial are:
1. There has to be a fixed number of trials.
2. Each trial must have the same two possible outcomes.
3. The outcomes of the trials have to be independent of one another.
4. The probability of each of the outcomes has to remain constant.

Conditions 3 and 4 are closely related, and it is sometimes difficult to articulate which is the essential reason the binomial is not appropriate. You don't need to worry about this, though, since identifying a problem with either is sufficient at this level.

Independence is a key condition. However, there are a number of situations where independence might be presumed to be present, but in fact is not.

The following are examples in which conditions 3 and 4 *do not* hold, and the binomial is *not* appropriate.

- Consider choosing five students from a group arriving at the recreational area at lunchtime. Why is it likely that things such as 'A-level subjects' and 'choice of meal' would not be independent?

  o *The lessons being taught before lunch will affect which students arrive together for lunch.*

- Consider a task to be performed repeatedly, for example an archer shooting at a target.

  o *The binomial conditions require that no learning takes place during the sequence of trials.*

- Consider a situation in which the underlying trial changes in some way. For example, consider the assumption that golfers have, individually, a fixed probability of getting a particular score on every hole.

  This assumption ignores various features of golf. For example:
  o *Each hole will have a varying degree of difficulty.*
  o *If one player in a match plays the hole particularly well, or badly, this may affect the strategy adopted by his opponents.*
  o *Independence between holes for the same player (or for the same hole but different players), is at best an approximation of reality.*

Independence does not apply if sampling without replacement is being undertaken. In a population of size N, the second observation comes from a reduced population of size $N - 1$, the third from a population of size $N - 2$, and so on. The larger the population is to start with, the smaller the error introduced by assuming that the probability remains fixed at the starting proportions.

> When n is large you can sometimes use the binomial distribution as an approximation, even if it isn't exactly right.

That is, if the conditions are not met exactly, it is still possible for the binomial to provide a useful model. The binomial is therefore used in a lot of real-life situations, where only a close approximation is required.

Exercise 7.3

1. For the following random variables state whether they can be modelled by a binomial. If they can, give the values of the parameters n and p. If they cannot, explain why.

   a) A die is thrown repeatedly until a 1 is seen. X = number of throws.

   b) A die is thrown 10 times. X = number of 1s seen.

   c) A bag has 25 red and 25 blue balls in it. 5 balls are taken out. X = the number of red balls taken out.

   d) X = number of boys in a family of 5 children.

   e) A pair of dice are rolled 25 times. X = number of times a double (that is, two 1s, two 2s, and so on) is thrown.

   f) A pair of dice are rolled 25 times. X = average of the sum of the numbers rolled.

2. a) If $X \sim B(30, 0.1)$, to find the probability that X is exactly

   i) 0   ii) 1

   iii) 2   iv) 3.

   b) State the mode of X.

   c) Give the mean and variance of X.

3. For the following situations, state what assumptions are needed if a binomial distribution is to be used to model them, and give the values of n and p that would be used. (You are not expected to do any calculations.)

   a) On average, a traffic warden gives a parking ticket to 8% of the cars he checks. One morning he checks 40 cars. How many tickets does he give out that morning?

   b) A box has 48 matches in it. On average, 2% of the matches made by that manufacturer do not strike properly. How many matches from the box will not strike properly?

   c) A bag has five red, three blue, and two green balls in it. Balls are taken out and the colour noted before the ball is returned. This is done 50 times. How many times was a blue ball taken out?

   d) A large drum has coloured balls in it; 50% are red, 30% are blue and 20% are green. 50 balls are removed and the number of blue balls is counted.

4. A vet thinks that the number of male puppies in litters of a given size will follow a binomial distribution with $p = 0.5$.

    a) In litters of six puppies, what would be the mean and variance of the number of males, if it is a binomial?

    The vet records the number of males in 82 litters of six puppies. The results are summarized in the table.

Number of males	0	1	2	3	4	5	6
Frequency	8	10	15	16	14	12	7

    b) Calculate the mean and variance of the number of males in litters of six puppies.

    c) Is the binomial a good model for the number of males in a litter of puppies?

Summary exercise 7

EXAM-STYLE QUESTION

1. A quality control agent tests sets of 10 components from a production line which is known to produce 98% good components.

    a) Find the probability that a set chosen at random is free from defects.

    b) If the quality control agent tests five sets before lunch, find the probability that four of these sets were free from defects.

    c) In one week, the agent tests 70 sets. On average, how many sets does she find which are not free from defects?

2. It is estimated that 4% of people have green eyes. In a random sample of size n, the expected number of people with green eyes is 5.

    a) Calculate the value of n.

    The expected number of people with green eyes in a second random sample is 3.

    b) Find the standard deviation of the number of people with green eyes in this second sample.

EXAM-STYLE QUESTION

3. A recent survey suggested that the proportion of 15-year-old girls who never consider their health when deciding what to eat is 0.1.

    Assuming that this figure is accurate, what is the probability that in a random sample of thirty 15-year-old girls the number who never consider their health when deciding what to eat is

    a) four or fewer

    b) exactly four?

4. Each evening Louise sets her alarm for 7:30 a.m.

She believes that the probability that she wakes before her alarm rings each morning is 0.4, and is independent from day to day.

a) Assuming that Louise's belief is correct, determine the probability that, during a week (five mornings), she wakes before her alarm rings

   i) on two or fewer mornings

   ii) on more than one but fewer than four mornings.

b) Assuming that Louise's belief is correct, calculate the probability that, during a four-week period, she wakes before her alarm rings on exactly seven mornings.

c) Assuming that Louise's belief is correct, calculate values for the mean and standard deviation of the number of mornings in a week when Louise wakes before her alarm rings.

d) During a 50-week period, Louise records, each week, the number of mornings on which she wakes before her alarm rings. The results are as follows.

Number of mornings	0	1	2	3	4	5
Frequency	10	11	9	9	6	5

   i) Calculate the mean and standard deviation of these data.

   ii) State, giving reasons, whether your answers to part (d)(i) support Louise's belief that the probability that she wakes before her alarm rings each morning is 0.4, and is independent from morning to morning.

5. The table below shows, for a particular population, the proportions of people in each of the four main blood groups.

Blood group	O	A	B	AB
Proportion	0.40	0.28	0.20	0.12

a) A random sample of 40 people is selected from the population. Determine the probability that the sample contains

   i) at most, ten people with blood group B

   ii) exactly five people with blood group AB

   iii) more than ten but fewer than 20 people with blood group O.

b) A random sample of 750 people is selected from the population.

Find the values for the mean and variance of the number of people in the sample with blood group A.

6. Ronnie and Parveen regularly play each other at tennis.

The probability that Ronnie wins any game is 0.3, and the outcome of each game is independent of the outcome of every other game.

a) Find the probability that, in a match of 15 games, Ronnie wins

   i) exactly four games

   ii) fewer than half of the games

   iii) exactly half of the games

   iv) more than one but fewer than six games.

Ronnie attends tennis coaching sessions for three months.

He then claims that the probability of him winning any game is 0.6, and that the outcome of each game is independent of the outcome of every other game.

b) i) Assuming this claim to be true, calculate the mean and standard deviation for the number of games won by Ronnie in a match of 15 games.

ii) To assess Ronnie's claim, Parveen keeps a record of the number of games won by Ronnie in a series of 10 matches, each of 15 games, with these results:

9 11 7 10 9 12 8 7 8 10

Calculate the mean and standard deviation of these values.
Hence comment on the validity of Ronnie's claim.

7. Plastic clothes pegs are made in several colours.

The number of blue pegs may be modelled by a binomial distribution with parameter p equal to 0.2.

The contents of packets of 40 pegs of several colours may be considered to be random samples.

a) Determine the probability that a packet contains

i) at most 10 blue pegs

ii) exactly 10 blue pegs

iii) more than 5 but fewer than 15 blue pegs.

b) Conn, a statistics student, claims to have counted the number of blue pegs in each of 100 packets of 40 pegs as part of a homework assignment. From his results these values are calculated:

Mean number of blue pegs per packet = 8.5

Variance of number of blue pegs per packet = 18.32

Comment on the validity of Conn's claim.

EXAM-STYLE QUESTION
8. Copies of an advertisement for a course in practical statistics are sent to mathematics teachers in a particular country. For each teacher who receives a copy, the probability of subsequently attending the course is 0.07.

18 teachers receive a copy of the advertisement.

What is the probability that the number who subsequently attend the course will be

a) two or fewer

b) exactly four?

9. a) State two assumptions of the binomial distribution.

b) A small garden decoration contains four electric light bulbs.
The probability that a bulb is faulty is 0.09.

i) What is the probability that exactly two bulbs are faulty?

ii) What is the probability that at least one bulb is faulty?

c) A set of outdoor lights consist of 20 light bulbs connected so that if at least one is faulty then none of the bulbs will light.

The probability that a bulb in the set is faulty is p.

Show that if $p = 0.034$ there is approximately a 50 : 50 chance that the set of lights does not light.

10. In a set of coloured beads used in costume jewellery, 10% are purple.

    a) Find the probability that in a string of 30 beads two or fewer beads are purple.

    b) Calculate the probability that in a string of 32 beads exactly two beads are purple.

    c) State one assumption that you have made in answering parts (a) and (b).

11. In a certain mountainous region in winter, the probability of more than 20 cm of snow falling on any particular day is 0.21.

    i) Find the probability that, in any 7-day period in winter, fewer than 5 days have more than 20 cm of snow falling. [3]

    ii) For four randomly chosen 7-day periods in winter, find the probability that exactly three of these periods will have at least 1 day with more than 20 cm of snow falling. [4]

    **Cambridge International AS and
    A Level Mathematics 9709,
    Paper 61 Q4 May/June 2012**

12. A company set up a display consisting of 20 fireworks. For each firework, the probability that it fails to work is 0.05, independently of other fireworks.

    i) Find the probability that more than 1 firework fails to work. [3]

    The 20 fireworks cost the company $24 each. 450 people pay the company $10 each to watch the display. If more than 1 firework fails to work they get their money back.

    ii) Calculate the expected profit for the company. [4]

    **Cambridge International AS and
    A Level Mathematics 9709,
    Paper 61 Q5 October/November 2012**

13. Biscuits are sold in packets of 18. There is a constant probability that any biscuit is broken, independently of other biscuits. The mean number of broken biscuits in a packet has been found to be 2.7. Find the probability that a packet contains between 2 and 4 (inclusive) broken biscuits. [4]

    **Cambridge International AS and
    A Level Mathematics 9709,
    Paper 61 Q1 May/June 2011**

Chapter summary

- The binomial probability distribution is defined as

$$P(X=r) = \binom{n}{r} p^r q^{n-r} \qquad r = 0, 1, 2, \ldots, n, \qquad q = 1 - p,$$

  where $\binom{n}{r} = {}^nC_r = \dfrac{n!}{r!(n-r)!}$.

- The parameters of the binomial distribution are n, the number of trials, and p, the probability of a 'success' on any one trial.

- The binomial distribution is often written $X \sim B(n, p)$.

- If $X \sim B(n, p)$, then mean, $E(X) = np$, and variance, $\text{Var}(X) = (\sigma^2) = npq$, where $q = 1 - p$.

- The conditions for the binomial are:

  1. There has to be a fixed number of trials.

  2. Each trial must have the same two possible outcomes.

  3. The outcomes of the trials have to be independent of one another.

  4. The probability of each of the outcomes has to remain constant.

- When n is large you can sometimes use the binomial distribution as an approximation, even if it isn't exactly right.

8 The normal distribution

Although the power of modern computers is now so great that a lot of modeling and simulations are now based on sampling from very large data sets rather than randomly generating from probability distribution like the normal, it is still a very powerful and widely used distribution because of the range of situation in which it provides a very good approximation to reality. Very many measurements of living things are approximately normal, for example in this rapeseed field the harvest per square metre, and the heights of individual rapeseed plants would be approximately normal distributions.

Objectives

After studying this chapter you should be able to:

- Understand the use of a normal distribution to model a continuous random variable, and use normal distribution tables;
- Solve problems concerning a variable X, where $X \sim N(\mu, \sigma^2)$, including:
  - finding the value of $P(X > x_1)$, or a related probability, given the values of x_1, μ, σ,
  - finding a relationship between x_1, μ and σ given the value of $P(X > x_1)$ or a related probability.

Before you start

You should know how to:

1. Solve linear simultaneous equations.
   e.g. Solve the simultaneous equations
   $$a + 1.25b = 18.25 \quad (1)$$
   $$a - 0.90b = 7.50 \quad (2)$$
   Subtract (2) from (1)
   $$2.15b = 10.75$$
   $$b = 5$$
   Substitute 5 for b in (1)
   $$a + 5 \times 1.25 = 18.25$$
   $$a = 12$$

2. Substitute values into simple equations.
   e.g. Find x, when $y = 5$, $a = 3$
   and $b = 2$ and $y = \dfrac{x-a}{h}$
   Substitute for y, a and b: $5 = \dfrac{x-3}{2}$
   $$10 = x - 3 \quad x = 13$$

Skills check:

1. Solve the simultaneous equations
   $$36 = \mu - 0.5\sigma$$
   $$46 = \mu + 0.5\sigma$$

2. Find p when $a = 16$, $b = 6$, $c = 2$
   given that $p = \dfrac{a-b}{c}$

8.1 Continuous probability distributions and the normal distribution

The **normal distribution** frequently occurs in the real world.

For example, the heights, or weights, of people often follow an approximate normal distribution.

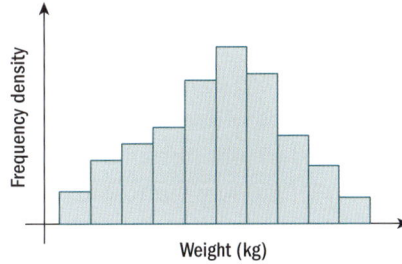

The dimensions of manufactured articles will usually follow a normal distribution as well.

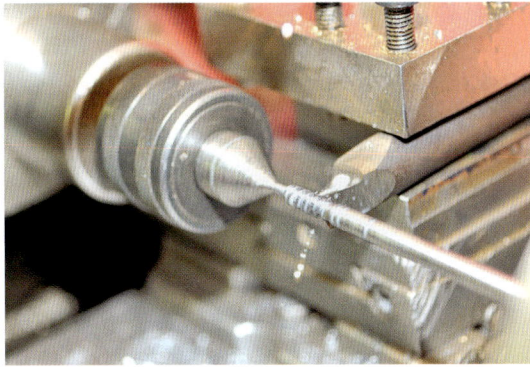

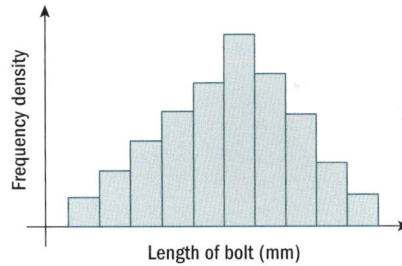

The normal distribution is a type of **continuous probability distribution**. This means that:

- It relates to a continuous variable (height, weight etc.)
- It describes the probability of this variable taking a particular range of values.

Generally, the normal distribution looks like this:

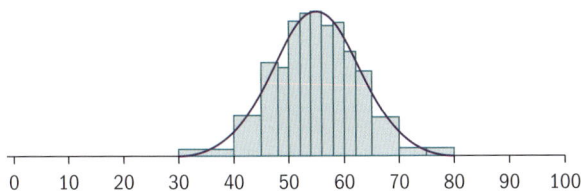

> **Did you know?**
> Physicists sometimes refer to this as the **Gaussian distribution**.

> The form of the function is
> $$f(x) = \frac{1}{\sigma\sqrt{2\pi}} e^{-\frac{1}{2}\left(\frac{x-\mu}{\sigma}\right)^2}.$$ This formula is not part of the Cambridge syllabus and so will not be required.

The idealised normal distribution has the following properties:

- It is symmetric
- It is infinite in both directions
- It has a single peak at the centre
- It is continuous
- 95% of values lie within approximately 2 standard deviations of the mean
- 99% lie within approximately 3 standard deviations of the mean.

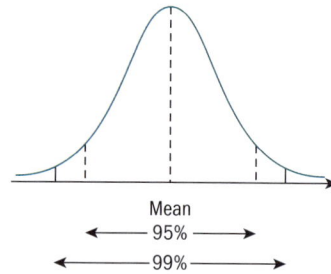

In practice, real-life variables usually do not match all these conditions perfectly and the situations will only approximate the normal distribution. For example, height and weight cannot take negative values, however a zero value may occur a long way out towards the tail of the distribution, as in the diagram above where the left hand end of the graph is at 20 and it is already almost invisible. If the value were not towards the tail, then the normal distribution would not be a reasonable approximation to use in that situation.

The follow will enable us to calculate probabilities:

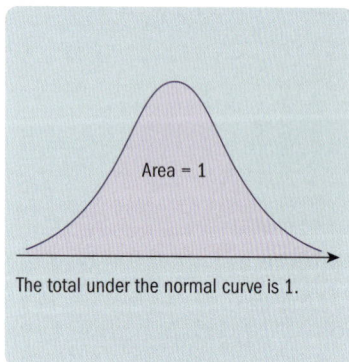

Area = 1

The total under the normal curve is 1.

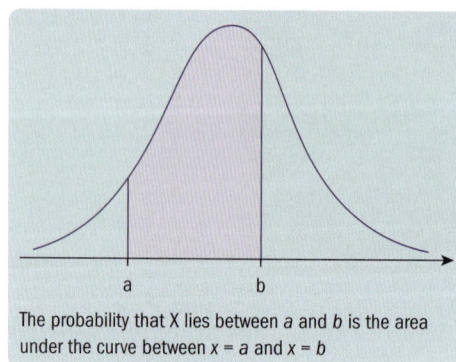

The probability that X lies between *a* and *b* is the area under the curve between $x = a$ and $x = b$

If you go on to study S2 then you will look at continuous distributions in more detail.

Standardised scores

The normal distribution allows us to make comparisons between individuals in a particular **normal population**. However it becomes harder with different normal populations – for example, we use different criteria to judge a 'tall man' than we use for a 'tall woman'.

In 2007 De-Fen Yao was 34 years old and 7ft 8" tall, making her the tallest woman ever. But how would she rank as a man in terms of height?

The tallest recorded man was Robert Pershing Wadlow at 8ft 11" tall – measured in 1940.

Yelena Isinbayeva is the female world record holder in the pole vault (5.06 metres) – is her achievement more or less exceptional than Usain Bolt's world record of 9.58 seconds for 100 metres?

One way to compare different normal populations is to standardise the scores – by looking at their distance from the mean, then dividing by the size of the standard deviation.

- To find the **standardised score, z**, from a raw score, x, use the conversion $z = \dfrac{x - \mu}{\sigma}$

  where μ is the mean and σ is the standard deviation of the raw scores.

This process allows comparison of performances of pupils in different subjects, or of athletes under different conditions. However, it is not the only process we may use to compare normal populations, and events like the decathlon and heptathlon use a points scoring system based on a different approach. See pages 190-191 for more on this.

Example 1

In her exams, Alexandra scores 75 in History and 87 in Maths. For the year group as a whole, History has a mean score of 63 on the examination with a standard deviation of 8, while Maths has a mean of 69 with a standard deviation of 15. Compare Alexandra's performance in these two subjects.

For History, $z = \dfrac{75 - 63}{8} = 1.5$

For Maths, $z = \dfrac{87 - 69}{15} = 1.2$

Alexandra's standardised score z is higher in History than in Maths, so there is reason to say that her performance is better in History than in Maths.

If we know corresponding points in two distributions, we should be able to work out an unknown mean or standard deviation.

Example 2

The mean height of a certain plant (A) is 67 cm, and the heights have a standard deviation of 5 cm. Another plant (B) has a mean height of 63 cm.
The same proportion of both types of plant is taller than 77 cm.
What is the standard deviation of the heights for plant B?

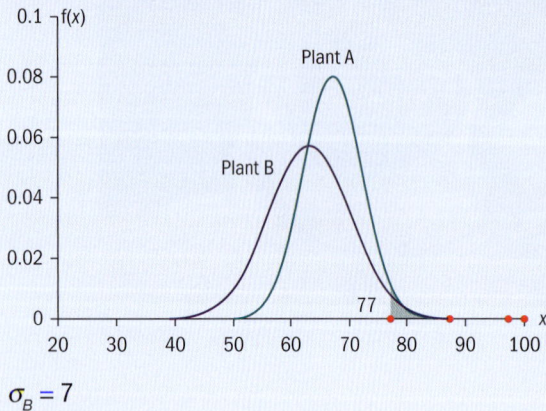

The vertical scale on this graph is the *relative frequency* you met in chapter 3.

77 cm is 10 cm above the mean for A, or 2 standard deviations. 77 cm is 14 cm above the mean for B and this must also be 2 standard deviations. Standard deviation of the heights for plant B.

$\sigma_B = 7$

Exercise 8.1

1. $\mu = 56 \quad \sigma = 7$

   a) Find the standardised score for a raw score of
      i) 70 ii) 52.5
      iii) 66.5 iv) 56.

   b) Find the raw score for a standardised score of
      i) 1.3 ii) −2.4
      iii) −0.4 iv) 2.0.

2. $\mu = 87$ $\sigma = 5$

    a) Find the standardised score for a raw score of

        i) 80   **ii)** 59   **iii)** 91.3   **iv)** 86.7.

    b) Find the raw score for a standardised score of

        i) 2.3   **ii)** −2.1   **iii)** −0.6   **iv)** 1.0.

3. $\mu = 3$ $\sigma = 12$

    a) Find the standardised score for a raw score of

        i) 15   **ii)** −3   **iii)** −27   **iv)** 5.8.

    b) Find the raw score for a standardised score of

        i) 0.7   **ii)** −1.3   **iii)** −0.2   **iv)** 1.8.

4. **a)** If $\mu = 64$, and 76 has a z score of 2, find σ.

    b) If $\sigma = 10$, and 43 has a z score of −1.6, find μ.

5. X has $\mu = 48$ and $\sigma = 10$. Y has mean 53.

    If the same proportion of X and Y are above 68, find the standard deviation of Y.

6. X has $\mu = 549$ and $\sigma = 34$. Y has a standard deviation of 47.

    If the same proportion of X and Y are above 600, find the mean of Y.

8.2 Standard normal distribution

- The normal distribution is written as $X \sim N(\mu, \sigma^2)$

This means 'X is distributed as a normal random variable with mean μ and variance σ^2'

Since all normal distributions are the same basic shape, we only need to have probabilities for one particular case to allow us to calculate probabilities for all cases.

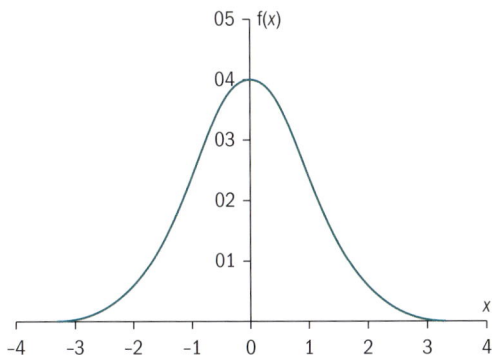

The **standard normal distribution** has mean 0 and variance 1.

The variable Z is often used for the standard normal distribution,

- For the standard normal distribution, $Z \sim N(0, 1^2)$ or $Z \sim N(0, 1)$.

By converting values to the standard normal distribution, you can use **probability tables** to calculate probabilities for any normal distribution.

> Be careful to distinguish between variance and standard deviation – if you are given $N(83, 16)$, then you need to use the standard deviation of 4.

z	0	1	2	3	4	5	6	7	8	9	1	2	3	4	5 ADD	6	7	8	9
0.0	0.5000	0.5040	0.5080	0.5120	0.5160	0.5199	0.5239	0.5279	0.5319	0.5359	4	8	12	16	20	24	28	32	36
0.1	0.5398	0.5438	0.5478	0.5517	0.5557	0.5596	0.5636	0.5675	0.5714	0.5753	4	8	12	16	20	24	28	32	36
0.2	0.5793	0.5832	0.5871	0.5910	0.5948	0.5987	0.6026	0.6064	0.6103	0.6141	4	8	12	15	19	23	27	31	35
0.3	0.6179	0.6217	0.6255	0.6293	0.6331	0.6368	0.6406	0.6443	0.6480	0.6517	4	7	11	15	19	22	26	30	34
0.4	0.6554	0.6591	0.6628	0.6664	0.6700	0.6736	0.6772	0.6808	0.6844	0.6879	4	7	11	14	18	22	25	29	32
0.5	0.6915	0.6950	0.6985	0.7019	0.7054	0.7088	0.7123	0.7157	0.7190	0.7224	3	7	10	14	17	20	24	27	31
0.6	0.7257	0.7291	0.7324	0.7357	0.7389	0.7422	0.7454	0.7486	0.7517	0.7549	3	7	10	13	16	19	23	26	29
0.7	0.7580	0.7611	0.7642	0.7673	0.7704	0.7734	0.7764	0.7794	0.7823	0.7852	3	6	9	12	15	18	21	24	27
0.8	0.7881	0.7910	0.7939	0.7967	0.7995	0.8023	0.8051	0.8078	0.8106	0.8133	3	5	8	11	14	16	19	22	25
0.9	0.8159	0.8186	0.8212	0.8238	0.8264	0.8289	0.8315	0.8340	0.8365	0.8389	3	5	8	10	13	15	18	20	23
1.0	0.8413	0.8438	0.8461	0.8485	0.8508	0.8531	0.8554	0.8577	0.8599	0.8621	2	5	7	9	12	14	16	19	21
1.1	0.8643	0.8665	0.8686	0.8708	0.8729	0.8749	0.8770	0.8790	0.8810	0.8830	2	4	6	8	10	12	14	16	18
1.2	0.8849	0.8869	0.8888	0.8907	0.8925	0.8944	0.8962	0.8980	0.8997	0.9015	2	4	6	7	9	11	13	15	17
1.3	0.9032	0.9049	0.9066	0.9082	0.9099	0.9115	0.9131	0.9147	0.9162	0.9177	2	3	5	6	8	10	11	13	14

The tables give you $P(Z \leq z)$, where z is any value from 0 to 3 (beyond that the probabilities have become so small they are negligible).

To find $P(Z \leq 1)$:

Locate z = 1.0 in the table
Go along to the column 0, and read off the value
$P(Z \leq 1) = 0.8413$

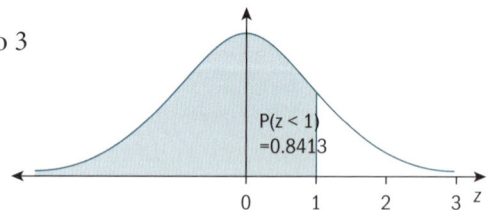

> Because the normal distribution is symmetrical, the tables only give you half the set of probabilities – the other half are identical.

So to find $P(Z \leq -1)$:

$P(Z \leq -1) = P(Z > 1)$
$\qquad = 1 - 0.8413$
$\qquad = 0.1587$

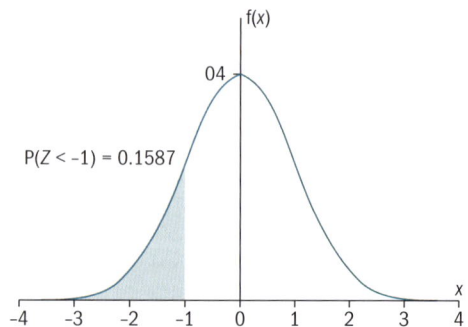

Because you always use the standard normal distribution to find probabilities it is worth having a special notation for this.

The tables are defined as the probability Z ≤ z, but this is the same as the probability Z < z (you will learn more about continuous distributions if you study S2).

We write $\Phi(z) = P(Z \le z)$.

So $\Phi(-k) = 1 - \Phi(k)$ by symmetry.

You often need to use more than one value in the tables:

$P(0.5 \le z \le 1.8) = \Phi(1.8) - \Phi(0.5) = 0.9641 - 0.6915 = 0.2726$

$P(-1 < z < 1.3) = \Phi(1.3) - \Phi(-1) = 0.9032 - (1 - 0.8413)$
$= 0.7445$

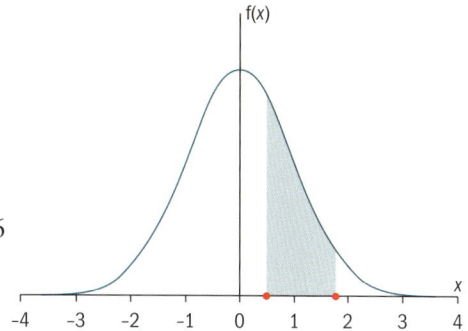

The tables allow us to give probabilities (correct to 4 d.p.) for z scores up to 2 decimal places, and estimates of probabilities for a 3rd decimal place in z.

It always helps to draw a clear sketch diagram.

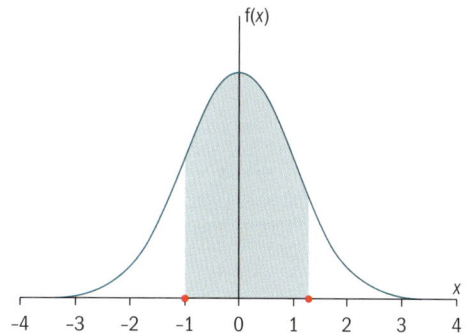

Example 3

If $Z \sim N(0, 1^2)$ find $P(Z < 0.247)$

z	0	1	2	3	4	5	6	7	8	9	1	2	3	4	5 ADD	6	7	8	9
0.0	0.5000	0.5040	0.5080	0.5120	0.5160	0.5199	0.5239	0.5279	0.5319	0.5359	4	8	12	16	20	24	28	32	36
0.1	0.5398	0.5438	0.5478	0.5517	0.5557	0.5596	0.5636	0.5675	0.5714	0.5753	4	8	12	16	20	24	28	32	36
0.2	0.5793	0.5832	0.5871	0.5910	0.5948	0.5987	0.6026	0.6064	0.6103	0.6141	4	8	12	15	19	23	27	31	35
0.3	0.6179	0.6217	0.6255	0.6293	0.6331	0.6368	0.6406	0.6443	0.6480	0.6517	4	7	11	15	19	22	26	30	34
0.4	0.6554	0.6591	0.6628	0.6664	0.6700	0.6736	0.6772	0.6808	0.6844	0.6879	4	7	11	14	18	22	25	29	32
0.5	0.6915	0.6950	0.6985	0.7019	0.7054	0.7088	0.7123	0.7157	0.7190	0.7224	3	7	10	14	17	20	24	27	31

To find the probability with a z score of 0.247 using the table, we find the row corresponding to $z = 0.2$. We then find the column headed '4' (corresponding to the second decimal place) and then the column headed '7' in the extra section to the right of the main table. These extra columns on the right-hand side give the *average* extra probability for the 3rd decimal place values at the top of the columns. We add the two digits found from the last column to the third decimal place of the number found in the previous column,

i.e. $0.5948 + 0.0027 = 0.5975$

and so,

$P(Z < 0.247) = 0.5975$

Example 4

If $Z \sim N(0, 1^2)$ find

a) $P(Z < 1.62)$ **b)** $P(Z > 0.76)$ **c)** $P(Z < -1.32)$ **d)** $P(-1.2 < Z < 1.7)$

a) $P(Z < 1.62) = \Phi(1.62) = 0.9474$

b) $P(Z > 0.76) = 1 - \Phi(0.76)$
$$= 1 - 0.7764 = 0.2236$$

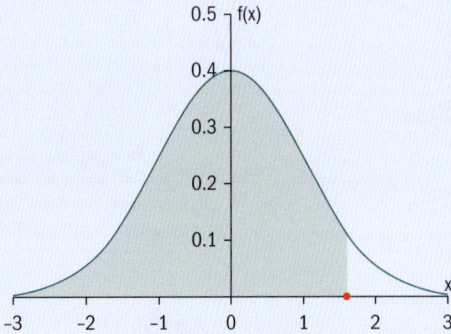

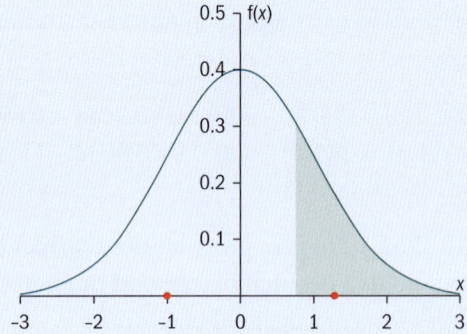

c) $P(Z < -1.32) = 1 - \Phi(1.32)$
$$= 1 - 0.9066$$
$$= 0.0934$$

d) $P(-1.2 < Z < 1.738) = \Phi(1.738) - \Phi(-1.2)$
$$= \Phi(1.738) - (1 - \Phi(1.2))$$
$$= 0.9589 - (1 - 0.8849)$$
$$= 0.8438$$

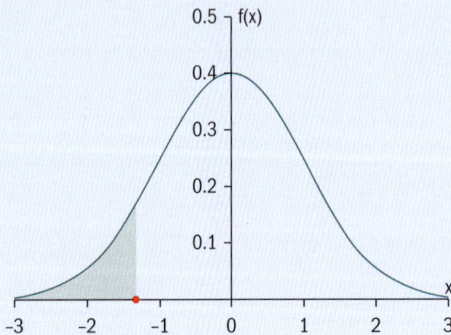

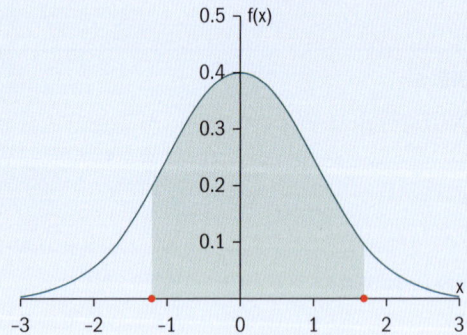

If you have a z score which is not listed, such as $z = \frac{5}{3}$, you should calculate it to 3 decimal places and use that value.

$\frac{5}{3} = 1.6666...... = 1.667$ (3 d.p.)

$\Phi(1.667) = 0.9522$

There is a second set of tables provided which give the exact z scores for a limited number of **tail probabilities**.
These enable us to work out the z score corresponding to a particular proportion.

For example, consider the z value corresponding to the top 5%.

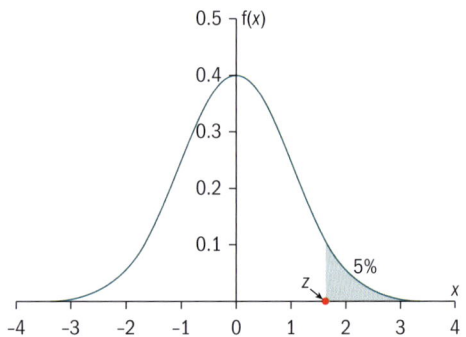

p	0.75	0.90	0.95	0.975	0.99	0.995	0.9975	0.999	0.9995
z	0.674	1.282	1.645	1.960	2.326	2.576	2.807	3.090	3.291

The table shows that $z = 1.645$.

If you need to find a z score corresponding to an unlisted probability, you can use the main table to find an approximation.

For example, consider the z value with $\Phi(z) = 0.5885$

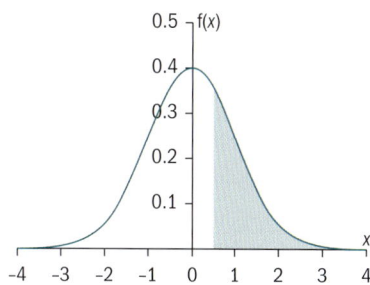

z	0	1	2	3	4	5	6	7	8	9	1	2	3	4	5	6	7	8	9
															ADD				
0.0	0.5000	0.5040	0.5080	0.5120	0.5160	0.5199	0.5239	0.5279	0.5319	0.5359	4	8	12	16	20	24	28	32	36
0.1	0.5398	0.5438	0.5478	0.5517	0.5557	0.5596	0.5636	0.5675	0.5714	0.5753	4	8	12	16	20	24	28	32	36
0.2	0.5793	0.5832	0.5871	0.5910	0.5948	0.5987	0.6026	0.6064	0.6103	0.6141	4	8	12	15	19	23	27	31	35
0.3	0.6179	0.6217	0.6255	0.6293	0.6331	0.6368	0.6406	0.6443	0.6480	0.6517	4	7	11	15	19	22	26	30	34
0.4	0.6554	0.6591	0.6628	0.6664	0.6700	0.6736	0.6772	0.6808	0.6844	0.6879	4	7	11	14	18	22	25	29	32

Even using the third decimal place corrections, there is no z value which gives exactly 0.5885. The nearest we can get is 0.5886 when z is 0.224.

Exercise 8.2

All of these questions relate to the Standard Normal Distribution i.e. $Z \sim N(0, 1^2)$.

1. Find   **a)** $\Phi(-0.06)$ **b)** $\Phi(2.63)$ **c)** $\Phi\left(\dfrac{4}{5}\right)$ **d)** $\Phi(2.5) - \Phi(1.2)$ **e)** $\Phi(1.43) - \Phi(-1.03)$

2. Find   **a)** $P(Z < 1.08)$ **b)** $P(Z > -0.3)$ **c)** $P(Z < -0.72)$ **d)** $P\left(\dfrac{5}{4} < Z < \dfrac{13}{6}\right)$

3. Find the probabilities shown by the shaded areas

a)

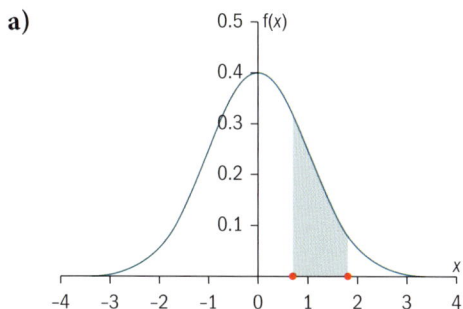

b)

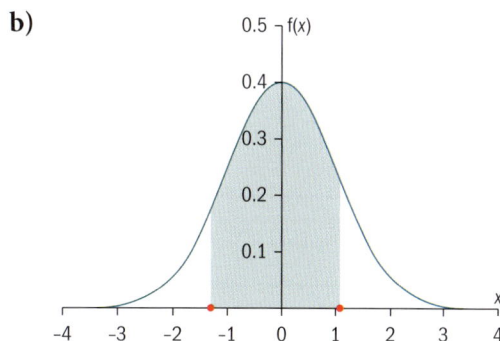

c)

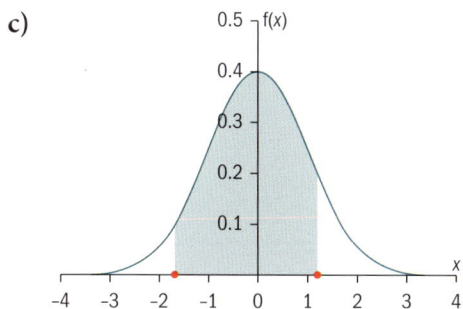

4. Find

    a) $P(|Z| < 1.8)$

    b) $P(|Z| > 0.72)$

    c) $P(Z < -2.8 \text{ or } Z > 2.1)$

    d) $P(Z < 1.4 \text{ or } X > 1.7)$

5. Find the z scores which cut off

    a) the top 10%

    b) the top 0.5%

    c) the bottom 2.5%

    d) the bottom 20%

    e) the top 6%

    f) the bottom 1.7%.

8.3 Calculating probabilities for the $N(\mu, \sigma^2)$ distribution

All normal distributions are essentially the same shape – they may have a different centre, or be more peaked, but they can all be standardised to the $N(0, 1)$ distribution.

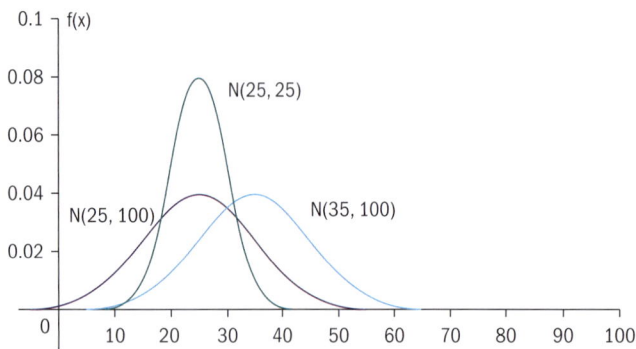

For a distribution $X \sim N(\mu, \sigma^2)$, we can find the probability of X taking a range of values through the following.

- First calculate z by using $z = \dfrac{x - \mu}{\sigma}$.
- Then use the table of probabilities to find $\Phi(z)$.
- Deduce the probability required, referring to a sketch.

Example 5

$X \sim N(2, 5^2)$.

Find **a)** $P(X < 7)$ **b)** $P(X > 11)$ **c)** $P(|X| < 3)$

      d) $P(|X - 2| < 6)$ **e)** x such that $P(X > x) = 0.05$

a) $P(X < 7)$

$$z = \frac{x - \mu}{\sigma} \Rightarrow z = \frac{7 - 2}{5} = 1$$

$$P(X < 7) = P(Z < 1) = \Phi(1) = 0.8413$$

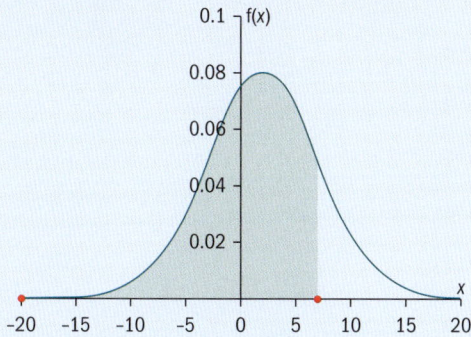

b) $P(X > 11)$

$$z = \frac{x - \mu}{\sigma} \Rightarrow z = \frac{11 - 2}{5} = 1.8$$

$$P(X > 11 = P(Z > 1.8) = -\Phi(1.8)$$
$$= 1 - 0.9641 = 0.0359$$

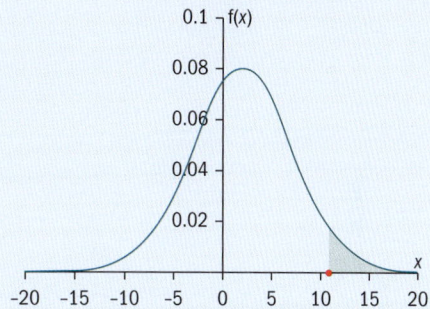

c) $P(|X| < 3)$

$|x| < 3 \Rightarrow -3 < x < 3$

$$z_1 = \frac{-3 - 2}{5} = -1 \qquad z_2 = \frac{3 - 2}{5} = 0.2$$

$$P(-3 < X < 3) = P(-1 < Z < 0.2)$$
$$= \Phi(0.2) - \Phi(-1) = 0.5793 - (1 - 0.8413) = 0.4206$$

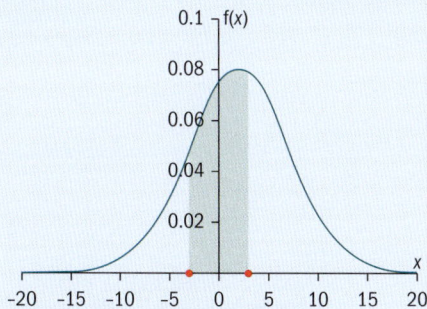

d) $P(|X - 2| < 6)$

$|x - 2| < 6 \Rightarrow -4 < x < 8$

$$z_1 = \frac{-4 - 2}{5} = -1.2 \qquad z_2 = \frac{8 - 2}{5} = 1.2$$

$$P(-4 < X < 8) = P(-1.2 < Z < 1.2) = \Phi(1.2)$$
$$- \Phi(-1.2) = 0.8849 - (1 - 0.8849) = 0.7698$$

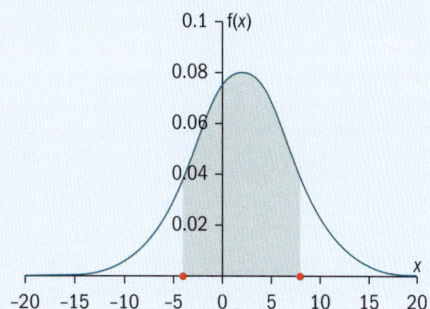

e) x such that $P(X > x) = 0.05$

$$\Phi(z) = 0.95 \Rightarrow z = 1.645$$

$$x = \mu + z\sigma \Rightarrow x = 2 + 1.645 \times 5 = 10.225$$

so $x = 10.2$ (to 3 s.f.)

> Use the table of percentage points to find $Z = 1.645$

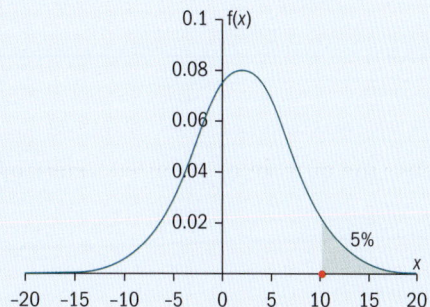

If we are given enough information, we can work out μ or σ or both.

When you are given the probability and have to work out x, use the table of percentage points instead, unless it is not given

Example 6

X is normally distributed such that $X \sim N(\mu, 36)$

Also, it is known that $P(X > 159.3) = 0.05$.

Calculate the value of μ correct to 1 decimal place.

$\Phi(z) = 0.95 \Rightarrow z = 1.645$

$x = \mu + z\sigma \Rightarrow 159.3 = \mu + 1.645 \times 6$

$\Rightarrow \mu = 159.3 - 1.645 \times 6 = 149.43$

so $\mu = 149.4$ to 1 d.p.

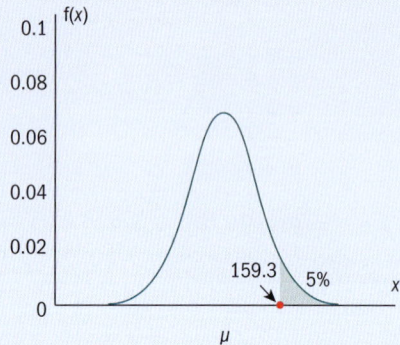

Example 7

$X \sim N(37.1, \sigma^2)$, $P(X > 51.3) = 0.04$

Calculate the value of σ.

$\Phi(z) = 0.96 \Rightarrow z = 1.751$

$x = \mu + z\sigma \Rightarrow 51.3 = 37.1 + 1.751 \times \sigma$

$\Rightarrow \sigma = \dfrac{51.3 - 37.1}{1.751} = 8.1096... = 8.11$ to 3 s.f.

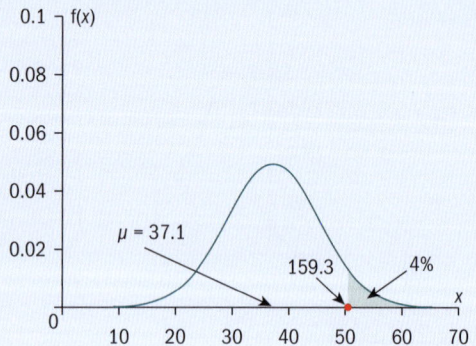

Calculating probabilities for the $N(\mu, \sigma^2)$ distribution

Example 8

$X \sim N(\mu, \sigma^2)$, $P(X < 37) = 0.1$, $P(X > 49.3) = 0.2$.

Calculate the values of μ and σ.

Hint: Here you need to form a pair of simultaneous equations.

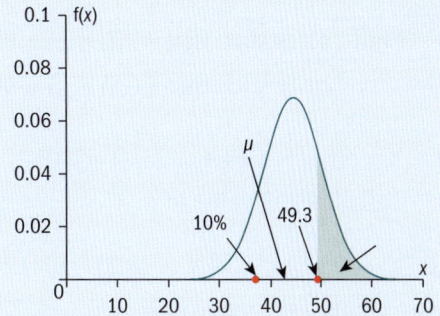

$\Phi(z_1) = 0.1 \Rightarrow z_1 = -1.282$

$\Phi(z_2) = 0.8 \Rightarrow z_2 = 0.842$

using $x = \mu + z\sigma \Rightarrow$

$49.3 = \mu + 0.842 \times \sigma$

$37 = \mu - 1.282 \times \sigma$

subtracting $\Rightarrow 12.3 = 2.124\sigma$

$\sigma = \dfrac{12.3}{2.124} = 5.793... = 5.79$ (3 s.f.)

$\mu = 37 + 1.282 \times 5.7909... = 44.424 ...$

$= 44.4$ (3 s.f.)

Note that 0.9 is given in the table of tail probabilities but 0.842 is not, so the z score of 0.842 was found in the main body of the table.

Exercise 8.3

1. $X \sim N(47, 5^2)$.

    Find **a)** $P(X < 56)$ **b)** $P(X > 51)$ **c)** $P(X < 42)$

2. $X \sim N(32, 16)$

    Find **a)** $P(X < 30)$ **b)** $P(X > 25)$ **c)** $P(X < 30.3)$

3. $X \sim N(4, 9)$

    Find **a)** $P(X < 10)$ **b)** $P(X > 5)$ **c)** $P(|X| < 3)$

4. $X \sim N(1750, 165^2)$

    Find **a)** $P(X < 1750)$ **b)** $P(X > 1780)$ **c)** $P(|X - 1750| < 165)$

5. $X \sim N(0, 20)$

    Find **a)** $P(X < 10)$ **b)** $P(X > 2.5)$ **c)** $P(-3 < X < 5)$

6. $X \sim N(25, 16)$

    Find **a)** x such that $P(X > x) = 0.05$ **b)** y such that $P(X < y) = 0.6$

7. $X \sim N(83.2, 4.5^2)$

    Find **a)** x such that $P(X > x) = 0.01$

           b) y such that $P(X < y) = 0.3$

           c) z such that $P(X < z) = 0.78$.

8. $X \sim N(0, 15)$

    Find **a)** x such that $P(|X| < x) = 0.8$

           b) y such that $P(|X| > y) = 0.6$.

9. $X \sim N(\mu, 5^2)$, $P(X > 23.4) = 0.05$.

    Calculate the value of μ.

10. $X \sim N(42, \sigma^2)$, $P(X > 48.3) = 0.01$.

    Calculate the value of σ.

11. $X \sim N(\mu, 16)$, $P(X > -3) = 0.98$.

    Calculate the value of μ.

12. $X \sim N(186, \sigma^2)$, $P(X < 193) = 0.92$.

    Calculate the value of σ.

13. $X \sim N(-32, \sigma^2)$, $P(X < -31.3) = 0.9$.

    Calculate the value of σ.

14. $X \sim N(\mu, \sigma^2)$, $P(X < 27) = 0.2$, $P(X > 35) = 0.3$.

    Calculate the values of μ and σ.

15. $X \sim N(\mu, \sigma^2)$, $P(X < 78) = 0.6$, $P(X > 89) = 0.2$.

    Calculate the values of μ and σ.

16. $X \sim N(\mu, \sigma^2)$, $P(X < 2.1) = 0.6$, $P(X < 2.7) = 0.7$.

    Calculate the values of μ and σ.

17. $X \sim N(\mu, \sigma^2)$, $P(X > 1056) = 0.6$, $P(X > 1132) = 0.2$.

    Calculate the values of μ and σ.

18. $X \sim N(\mu, \sigma^2)$, $P(X < 47.3) = 0.5$, $P(X > 52) = 0.2$.

    Calculate the values of μ and σ.

8.4 Using the normal distribution

The introduction to this chapter refers to the real-life uses of the normal distribution. This section ties together the techniques covered so far and describes how it is used in practice.

Example 9

The lengths of steel girders produced in a factory are normally distributed with a mean length of 12.5 m and a variance of 0.0004 m².

Girders need to be between 12.47 and 12.53 metres to be used in construction.

a) Find the proportion of girders which cannot be used for construction.

Girders are extremely expensive to produce, and the company is not happy with this level of wastage. A new machine is installed which reduces the variance of the production to 0.0002 m² while maintaining a Normal distribution with the same mean as before.

b) Find the proportion of girders produced by the new machine which cannot be used for construction.

a) The standard deviation of production is 0.02 m, or 2 cm.

$X \sim N(12.5, 0.02^2)$

$z_1 = \dfrac{12.47 - 12.5}{0.02} = -1.5; \quad z_2 = \dfrac{12.53 - 12.5}{0.02} = 1.5$

$P(12.47 < X < 12.53) = P(-1.5 < Z < 1.5) = \Phi(1.5) - \Phi(-1.5)$

$= 0.9332 - (1 - 0.9332) = 0.8664$

13.4% of the girders cannot be used.

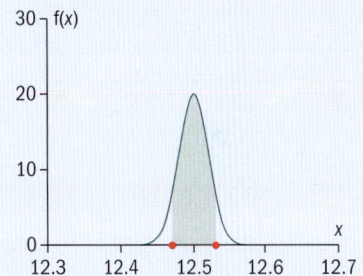

b) The standard deviation of production is now 0.01 m, or 1 cm.

$X \sim N(12.5, 0.01^2)$

$z_1 = \dfrac{12.47 - 12.5}{0.01} = -3; \quad z_2 = \dfrac{12.53 - 12.5}{0.01} = 3$

$P(12.47 < X < 12.53) = P(-3 < Z < 3) = \Phi(3) - \Phi(-3)$

$= 0.9987 - (1 - 0.9987) = 0.9974$

Now only 0.26% of the girders from the new machine cannot be used.

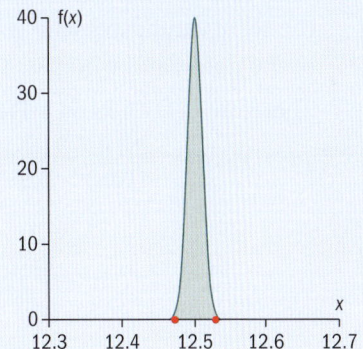

If we have historical exam performance data, we can predict the proportion of students that will get a particular grade.

Example 10

In an examination the marks are normally distributed with mean 60.7 and a standard deviation of 12.3.

a) A candidate needs a mark of at least 40 to pass.

What percentage of candidates fail?

b) The board awards distinctions to the best 10% of the candidates.

What is the least mark a candidate will need to get a distinction?

c) The list of passing candidates is published the day before the list of distinctions. What is the probability that a candidate who has passed will have got a distinction?

a) $X \sim N(60.7, 12.3^2)$

$$z = \frac{40 - 60.7}{12.3} = -1.6829\ldots$$

$P(40 < X) = P(-1.6829\ldots < Z) = 1 - \Phi(1.65)$

$$= 1 - 0.9537 = 0.0463$$

so 4.63% of the candidates fail.

b) $X \sim N(60.7, 12.3^2)$

$\Phi(z) = 0.9 \Rightarrow z = 1.2816$

$x = 60.7 + 1.2816 \times 12.3 = 76.46368$

so a candidate needs to score at least 77 to get a distinction.

c) This is a conditional probability, but a special sort – the set of candidates who get a distinction are a subset of the candidates who pass the exam.

Therefore $P(D \cap Pass) = P(D)$, and

$$P(D \mid Pass) = \frac{P(D)}{P(Pass)} = \frac{0.1}{0.9505} = 0.1052\ldots = 10.5\%$$

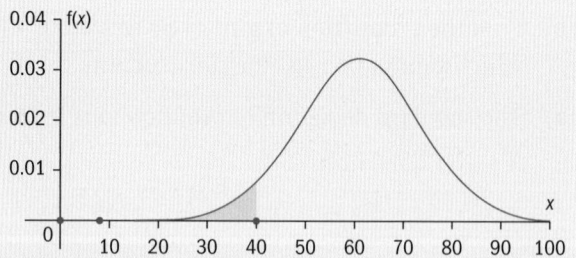

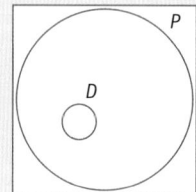

Companies producing packaged foods are usually required by law to state the estimated weight of the contents. They need to be able to show that it is rare for their products to contain less than what it says on the packet. However, all production processes are subject to variation.

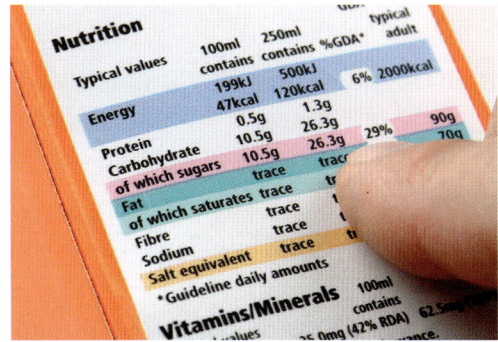

Example 11

A machine pours melted chocolate into moulds.
The standard deviation of the amount it pours is 0.7 grams, and the mean amount can be set on the machine.
The amount poured may be assumed to follow a normal distribution.
The machine is to produce bars whose label says 50 g of chocolate.
The company's lawyers want to have no more than 0.5% of bars containing less than the advertised weight.
What should the mean be set at?

0. 5% is equivalent to a probability of 0.005

$\Phi(z) = 0.005 \Rightarrow z = -2.5758$

$x = \mu + zs \Rightarrow 50 = \mu - 2.5758 \times 0.7$

$\Rightarrow \mu = 50 + 2.5758 \times 0.7 = 51.80306 = 51.8$ (3 s.f.)

So the mean should be set at 51.8 g.

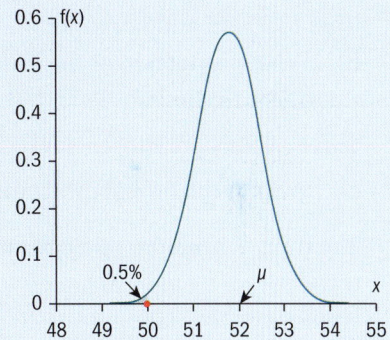

Example 12

An airline does a survey of the weights of adult passengers travelling on its flights.
It finds that 5% weigh more than 84.3 kg and 2% weigh less than 57.2 kg.
Assuming that the weights of adult passengers on its flights are normally distributed, find the mean and standard deviation of the weights.

$\Phi(z_1) = 0.02 \Rightarrow z_1 = -2.054 \quad \Phi(z_2) = 0.95 \Rightarrow z_2 = 1.6449$

using $x = \mu + zs \Rightarrow$

$84.3 = \mu + 1.6449 \times \sigma$

$57.2 = \mu - 2.054 \times \sigma$

subtracting $\Rightarrow 27.1 = 3.6989\sigma$

$\sigma = \dfrac{27.1}{3.6989} = 7.3265... = 7.33$ (3 s.f.)

$\mu = 57.2 + 2.054 \times 7.3265... = 72.2486... = 72.2$ (3 s.f.)

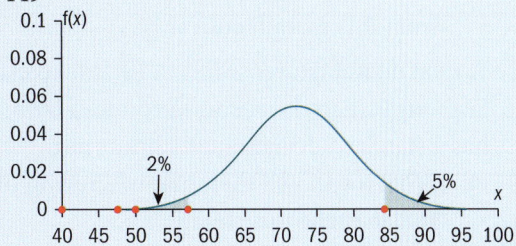

Exercise 8.4

1. IQ scores are normally distributed and are measured on a scale which has a mean of 100 and a standard deviation of 15. Find the IQ, X, which is exceed by only 5% of the population.

2. The lengths, L metres, that Philippe jumps in the triple jump event may be modelled by a normal distribution with mean 15.85 m and variance 0.36 m².

   a) For a jump chosen at random, find the probability that he jumps at least 16.2 m.

   In a competition, there is a qualifying distance of 16.2 m.

   b) Find the probability that he jumps over 16.5 m, given that he makes the qualifying distance with the jump.

3. The diameter of pistons produced in a factory are D cm, where $D \sim N(\mu, 0.12^2)$.

   a) Given that 5% of the pistons have a diameter less than 13.20 cm, show that $\mu = 13.4$.

   The tolerance specified for the pistons is that the diameter needs to be at least 13.35 cm and not more than 13.5 cm.

   b) What proportion of the production in the factory meets these tolerance limits?

   Three pistons are chosen at random.

   c) What is the probability that none of them meet the tolerance limits?

4. The time, T minutes, taken for the employees of a large firm to travel to work can be assumed to be normally distributed. Of the employees, 10% take at least 40 minutes to travel and 0.5% take less than 8 minutes.

   Find the mean and standard deviation of T.

5. The operational life of batteries produced by one manufacturer (A) is normally distributed with mean 43 hours and standard deviation 4 hours.

   a) Find the probability that a randomly chosen battery operates for more than 40 hours.

   b) What length of operational life is exceeded by 10% of batteries from this manufacturer?

   A rival manufacturer wants to be able to claim that 95% of its batteries last longer than the average produced by manufacturer (A). Their process has a standard deviation of 4 hours also.

   c) What mean will their manufacturing process need to have to be able to make this claim?

6. A machine fills tins with an amount of liquid which is normally distributed, with mean 330 ml, and standard deviation of 8 ml.

    a) Find the probability that a tin contains less than 320 ml.

    b) Find the probability that a tin contains between 320 and 345 ml.

    c) What volume is exceeded by 5% of the tins?

    Another machine also distributes the same liquid with amounts which follow a normal distribution.

    d) If the standard deviation is still 8 ml, what should the mean amount be set to if only 5% is to be below 320 ml?

    e) If the mean is to stay at 330 ml, and only 5% is to be below 320 ml, what would the standard deviation have to be.

7. The weights of eggs from one farm are normally distributed with mean 53 g and standard deviation 4 g.

    a) What is the probability that an egg from this farm weighs more than 56 grams?

    Eggs which weigh less than 48 grams are removed and not sold.

    b) What is the probability that an egg sold from this farm weighs less than 48 grams?

8. A company employs a large number of administrative staff. When the company wants to employ new staff, candidates are given a standard task to complete, and the time they take to complete the task is recorded. It is observed that the times taken by candidates are normally distributed with mean 360 seconds and standard deviation 75 seconds.

    a) i)   What proportion of the applicants take longer than 450 seconds?

        ii) What proportion of the applicants take between 210 seconds and 450 seconds?

        iii) What time is exceeded by 5% of the candidates?

    Candidates who take longer than 450 seconds are automatically rejected and those who take less than 210 seconds are automatically accepted. The remainder are interviewed.

    b) What is the median time taken by those applicants who are interviewed?

9. The volume of oil poured into cans in a production process is known to have a standard deviation of 2.6 ml.
    Of the cans, 22% contain less than 660 ml.

    a) Calculate the mean volume of oil in the cans.

    b) Calculate the proportion of cans which contain less than 650 ml.

10. The screws produced by a company have mean length 17.9 mm and variance 0.04 mm².

    a) What proportion of the screws are less than 17.5 mm.

    Any screws which are less than 17.5 mm or greater than 18.4 mm cannot be sold.

    b) In a batch of 1000 screws, how many would you expect to be rejected?

    The mean length can be adjusted by a setting on the machine.

    c) At what should the mean be set in order to maximise the proportion of screws satisfying the specifications?

    If a screw is too short it has to be scrapped, but if it is too long it can be filed to make it usable. The mean is set at 17.9 mm.

    d) A screw is chosen at random and found to be long enough not to be scrapped. What is the probability that it will need to be filed before it can be sold?

Summary exercise 8

1. IQs are measured on a scale with a mean of 100 and standard deviation 15. Assuming that IQs can be modelled by a normal random variable,

    find

    a) $P(Y > 130)$

    b) $P(73 \leq Y \leq 91)$

    c) the value of k, to 1 decimal place, such that $P(Y \leq k) = 0.2$.

2. The performances in the 100 metre sprint of a group of decathletes are modelled by a normal distribution with mean 11.46 seconds and a standard deviation of 0.32 seconds. The performances of this group of decathletes in the shot put are modelled by a normal distribution with mean 11.62 m and standard deviation 0.73 m.

    a) Find the probability that a randomly chosen athlete runs the 100 metre sprint faster than 10.8 seconds,

    b) Find the probability that a randomly chosen athlete puts the shot further than 12.8 m.

    c) Assuming that for these decathletes the performances in the two events are independent, find the probability that a randomly chosen athlete runs the 100 metre sprint faster than 10.8 seconds and puts the shot further than 12.8 m.

    d) Comment on the assumption that the performances in the two events are independent.

3. Bottles of spring water have a stated volume of 330 ml. The bottles are filled by a machine, which dispenses a volume which is normally distributed with mean 335 ml and standard deviation 5 ml.

   a) Find the probability of a bottle containing less than the stated volume.

   The bottles have a capacity of 345 ml. Any time that the machine dispenses more than this, the extra spring water spills out.

   b) Find the probability of a bottle containing less than the stated volume, given that it didn't overflow.

   A new machine is installed. Only 0.5% of bottles overflow when the mean amount dispensed by the new machine is set to 335 ml.

   c) Find the standard deviation of the amount of spring water dispensed by the new machine.

4. A random variable X has a normal distribution.

   a) Describe two features of the distribution of X.

   A company produces batteries which have lifespans that are normally distributed. Only 2% of the batteries have a lifespan less than 230 hours and 5% have a lifespan greater than 340 hours.

   b) Determine the mean and standard deviation of the lifespans of the batteries.

   The company gives a warranty of 250 hours on the batteries. They make a profit of £750 on each battery that they sell. Replacing a battery under warranty costs the company £1250.

   c) Find the average profit they make on the sale of 100 batteries after any replacements under warranty have been made.

5. The random variable X is normally distributed with mean 253 and variance 121. Find

   a) $P(X < 240)$

   b) $P(245 < X < 275)$.

   It is known that $P(a \le X) = 0.13$.

   c) Find the value of a.

6. The length of time it takes an examiner to mark an examination script may be modelled by a normal distribution with mean 8 minutes and standard deviation 90 seconds.

   Find the time, t minutes, such that one script in six will take the examiner longer than t minutes to mark.

7. An examination in Statistics consists of a written paper and a project.

   Marks for the written paper, E, may be modelled by a normal distribution with mean 62 and standard deviation 9. Marks for the project, F, may be modelled by a normal distribution with mean 70 and standard deviation 6.

   a) Find $P(E > 80)$

   b) Find p such that $P(F > p) = P(E > 80)$

   A distinction on the examination requires at least 75 on both the written paper and the project.

   c) Find the probability that a candidate gets a distinction, assuming that the performance on the written paper and the project are independent of one another.

   d) Comment on the assumption of independence on part (c).

8. The random variable $X \sim N(\mu, \sigma^2)$.

   It is known that
   $P(X \leq 69) = 0.0228$ and $P(X \geq 95) = 0.1056$

   a) i) Show that the value of σ is 8

   ii) Find the value of μ.

   b) Find $P(71 \leq X \leq 81)$.

9. An electronic component has a useful lifespan which can be modelled by a normal distribution with mean 8000 hours and a standard deviation of 400 hours.

   a) Calculate the probability that a randomly selected component will last

   i) less than 7700 hours

   ii) between 7500 and 8300 hours

   iii) at least a year if it is installed on 1st January 2007.

   The components cost the manufacturer $215 to produce. The manufacturer offers an optional guarantee at an extra cost of $50.
   The terms of the guarantee are that the manufacturer will replace the component free if its useful lifespan is less than 7000 hours, and at a cost of $75 to the customer if its lifespan is between 7000 and 7500 hours.

   b) Calculate the expected profit or loss on each guarantee sold by the manufacturer.

10. i) Give an example of a variable in real life which could be modelled by a normal distribution. [1]

    ii) The random variable X is normally distributed with mean μ and variance 21.0. Given that $P(X > 10.0) = 0.7389$, find the value of μ. [3]

    iii) If 300 observations are taken at random from the distribution in part (ii), estimate how many of these would be greater than 22.0. [4]

    Cambridge International AS and A Level Mathematics 9709, Paper 6 Q5 October/November 2006

11. Tyre pressures on a certain types of car independently follow a normal distribution with mean 1.9 bars and standard deviation 0.15 bars.

    i) Find the probability that all four tyres on a car of this type have pressures between 1.82 bars and 1.92 bars. [5]

    ii) Safety regulations state that the pressures must be between $1.9 - b$ bars and $1.9 + b$ bars. It is known that 80% of tyres are within these safety limits. Find the safety limits. [3]

    Cambridge International AS and A Level Mathematics 9709, Paper 6 Q6 May/June 2005

12. a) The random variable X is normally distributed with mean μ and standard deviation σ. It is given that $3\mu = 7\sigma^2$ and that $P(X > 2\mu) = 0.1016$. Find μ and σ. [4]

    b) It is given that $Y \sim N(33, 21)$. Find the value of a given that $P(33 - a < Y < 33 + a) = 0.5$. [4]

    Cambridge International AS and A Level Mathematics 9709, Paper 61 Q5 May/June 2011

13. The lengths of body feathers of a particular species of bird are modelled by a normal distribution. A researcher measures the lengths of a random sample of 600 body feathers from birds of this species and finds that 63 are less than 6 cm long and 155 are more than 12 cm long.

i) Find estimates of the mean and standard deviation of the lengths of body feathers of birds of this species. [5]

ii) In a random sample of 1000 body feathers from birds of this species, how many would the researcher expect to find with lengths more than 1 standard deviation from the mean? [4]

Cambridge International AS and A Level Mathematics 9709, Paper 61 Q6 May/June 2012

14.

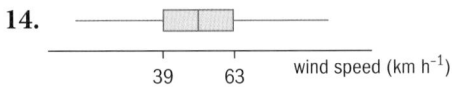

wind speed (km h⁻¹): 39, 63

Measurements of wind speed on a certain island were taken over a period of one year. A box-and-whisker plot of the data obtained is displayed above, and the values of the quartiles are as shown. It is suggested that wind speed can be modelled approximately by a normal distribution with mean μ km h^{-1} and standard deviation σ km h^{-1}.

i) Estimate the value of μ. [1]

ii) Estimate the value of σ. [3]

Cambridge International AS and A Level Mathematics 9709, Paper 62 Q1 October/November 2009

15. The lengths of fish of a certain type have a normal distribution with mean 38 cm. It is found that 5% of the fish are longer than 50 cm.

i) Find the standard deviation. [3]

ii) When fish are chosen for sale, those shorter than 30 cm are rejected. Find the proportion of fish rejected. [3]

iii) 9 fish are chosen at random. Find the probability that at least one of them is longer than 50 cm [2]

Cambridge International AS and A Level Mathematics 9709, Paper 6 Q3 May/June 2006

16. a) The random variable Y is normally distributed with positive mean μ and standard deviation $\frac{1}{2}\mu$. Find the probability that a randomly chosen value of y is negative [3]

b) The weights of bags of rice are normally distributed with mean 2.04 kg and standard deviation σ kg. In a random sample of 8000 such bags, 253 weighted over 2.1 kg. Find the value of σ [4]

Cambridge International AS and A Level Mathematics 9709, Paper 61 Q4, May/June 2013

17. The probability that New Year's Day is on a Saturday in a randomly chosen year is $\frac{1}{7}$.

i) 15 years are chosen randomly. Find the probability that at least 3 of these years have New Year's Day on a Saturday. [4]

ii) 56 years are chosen randomly. Use a suitable approximation to find the probability that more than 7 of these years have New Year's Day on a Saturday. [5]

Cambridge International AS and A Level Mathematics 9709, Paper 6 Q6 May/June 2007

Chapter summary

Continuous probability distributions and the normal distribution

- The normal distribution is continuous, symmetric, infinite in both directions and has a single peak at the centre.

  - ➢ 95% of values lie within approximately 2 standard deviations of the mean,

  - ➢ 99% lie within approximately 3 standard deviations of the mean.

- To find the standardised score, z, from a raw score, x, use the conversion $z = \dfrac{x - \mu}{\sigma}$ where μ is the mean and σ is the standard deviation of the raw scores. $x = \mu + z\sigma$ can be used to convert standardised scores back to raw scores.

- The probability tables for the N(0, 1) distribution give the cumulative probability $\Phi(z)$ for non-negative values of z. The symmetry of the distribution allows the values of $\Phi(z)$ for negative z to be deduced from these. All probabilities can then be worked out using one or two values from the tables.

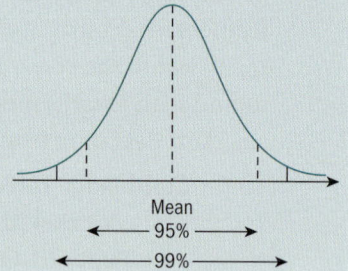

Standard normal distribution

- The normal distribution is written as $X \sim N(\mu, \sigma_2)$
- The **standard normal distribution** has mean 0 and variance 1.

Calculating probabilities for the N(μ, σ^2) distribution

- Calculating probabilities for the N(μ, σ^2) distribution is done by standardising the scores and using the standard normal distribution tables.
- Calculating an unknown mean and/or standard deviation is done by constructing one or two equations involving the unknowns from the probability information provided and then solving for the unknown(s).

List of formulae and tables of the normal distribution

If Z has a normal distribution with mean 0 and variance 1 then, for each volume of z, the table gives the values of $\Phi(z)$, where $\Phi(z) = P(Z \leq z)$.

For negative values of z use $\Phi(-z) = 1 - \Phi(z)$.

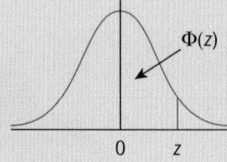

z	0	1	2	3	4	5	6	7	8	9	1	2	3	4	5 ADD	6	7	8	9
0.0	0.5000	0.5040	0.5080	0.5120	0.5160	0.5199	0.5239	0.5279	0.5319	0.5359	4	8	12	16	20	24	28	32	36
0.1	0.5398	0.5438	0.5478	0.5517	0.5557	0.5596	0.5636	0.5675	0.5714	0.5753	4	8	12	16	20	24	28	32	36
0.2	0.5793	0.5832	0.5871	0.5910	0.5948	0.5987	0.6026	0.6064	0.6103	0.6141	4	8	12	15	19	23	27	31	35
0.3	0.6179	0.6217	0.6255	0.6293	0.6331	0.6368	0.6406	0.6443	0.6480	0.6517	4	7	11	15	19	22	26	30	34
0.4	0.6554	0.6591	0.6628	0.6664	0.6700	0.6736	0.6772	0.6808	0.6844	0.6879	4	7	11	14	18	22	25	29	32
0.5	0.6915	0.6950	0.6985	0.7019	0.7054	0.7088	0.7123	0.7157	0.7190	0.7224	3	7	10	14	17	20	24	27	31
0.6	0.7257	0.7291	0.7324	0.7357	0.7389	0.7422	0.7454	0.7486	0.7517	0.7549	3	7	10	13	16	19	23	26	29
0.7	0.7580	0.7611	0.7642	0.7673	0.7704	0.7734	0.7764	0.7794	0.7823	0.7852	3	6	9	12	15	18	21	24	27
0.8	0.7881	0.7910	0.7939	0.7967	0.7995	0.8023	0.8051	0.8078	0.8106	0.8133	3	5	8	11	14	16	19	22	25
0.9	0.8159	0.8186	0.8212	0.8238	0.8264	0.8289	0.8315	0.8340	0.8365	0.8389	3	5	8	10	13	15	18	20	23
1.0	0.8413	0.8438	0.8461	0.8485	0.8508	0.8531	0.8554	0.8577	0.8599	0.8621	2	5	7	9	12	14	16	19	21
1.1	0.8643	0.8665	0.8686	0.8708	0.8729	0.8749	0.8770	0.8790	0.8810	0.8830	2	4	6	8	10	12	14	16	18
1.2	0.8849	0.8869	0.8888	0.8907	0.8925	0.8944	0.8962	0.8980	0.8997	0.9015	2	4	6	7	9	11	13	15	17
1.3	0.9032	0.9049	0.9066	0.9082	0.9099	0.9115	0.9131	0.9147	0.9162	0.9177	2	3	5	6	8	10	11	13	14
1.4	0.9192	0.9207	0.9222	0.9236	0.9251	0.9265	0.9279	0.9292	0.9306	0.9319	1	3	4	6	7	8	10	11	13
1.5	0.9332	0.9345	0.9357	0.9370	0.9382	0.9394	0.9406	0.9418	0.9429	0.9441	1	2	4	5	6	7	8	10	11
1.6	0.9452	0.9463	0.9474	0.9484	0.9495	0.9505	0.9515	0.9525	0.9535	0.9545	1	2	3	4	5	6	7	8	9
1.7	0.9554	0.9564	0.9573	0.9582	0.9591	0.9599	0.9608	0.9616	0.9625	0.9633	1	2	3	4	4	5	6	7	8
1.8	0.9641	0.9649	0.9656	0.9664	0.9671	0.9678	0.9686	0.9693	0.9699	0.9706	1	1	2	3	4	4	5	6	6
1.9	0.9713	0.9719	0.9726	0.9732	0.9738	0.9744	0.9750	0.9756	0.9761	0.9767	1	1	2	2	3	4	4	5	5
2.0	0.9772	0.9778	0.9783	0.9788	0.9793	0.9798	0.9803	0.9808	0.9812	0.9817	0	1	1	2	2	3	3	4	4
2.1	0.9821	0.9826	0.9830	0.9834	0.9838	0.9842	0.9846	0.9850	0.9854	0.9857	0	1	1	2	2	2	3	3	4
2.2	0.9861	0.9864	0.9868	0.9871	0.9875	0.9878	0.9881	0.9884	0.9887	0.9890	0	1	1	1	2	2	2	3	3
2.3	0.9893	0.9896	0.9898	0.9901	0.9904	0.9906	0.9909	0.9911	0.9913	0.9916	0	1	1	1	1	2	2	2	2
2.4	0.9918	0.9920	0.9922	0.9925	0.9927	0.9929	0.9931	0.9932	0.9934	0.9936	0	0	1	1	1	1	1	2	2
2.5	0.9938	0.9940	0.9941	0.9943	0.9945	0.9946	0.9948	0.9949	0.9951	0.9952	0	0	0	1	1	1	1	1	1
2.6	0.9953	0.9955	0.9956	0.9957	0.9959	0.9960	0.9961	0.9962	0.9963	0.9964	0	0	0	0	1	1	1	1	1
2.7	0.9965	0.9966	0.9967	0.9968	0.9969	0.9970	0.9971	0.9972	0.9973	0.9974	0	0	0	0	0	1	1	1	1
2.8	0.9974	0.9975	0.9976	0.9977	0.9977	0.9978	0.9979	0.9979	0.9980	0.9981	0	0	0	0	0	0	0	1	1
2.9	0.9981	0.9982	0.9982	0.9983	0.9984	0.9984	0.9985	0.9985	0.9986	0.9986	0	0	0	0	0	0	0	0	0

Critical values for the normal distribution

If Z has a normal distribution with mean 0 and variance 1 then, for each value of p, the table gives the value of z such that $P(Z \leq z) = p$.

p	0.75	0.90	0.95	0.975	0.99	0.995	0.9975	0.999	0.9995
z	0.674	1.282	1.645	1.960	2.326	2.576	2.807	3.090	3.291

This is a particular case of a very important theorem – The central limit theorem which says that the distribution of means of samples of size n from any underlying distribution will be approximately normal if n is large enough. The fundamental stability and consistency of behavior of large numbers of individually random events is the basis for some financial products and of insurance markets.

Objectives

After studying this chapter you should be able to:

- Recall conditions under which the normal distribution can be used as an approximation to the binomial distribution (n large enough to ensure that $np > 5$ and $nq > 5$), and use this approximation, with a continuity correction, in solving problems.

Before you start

You should know how to:

1. Find the mean and variance of a binomial distribution, e.g. If $X \sim B(12, 0.6)$, find the mean and variance of X.

   Mean $= 12 \times 0.6 = 7.2$;
   Variance $= 12 \times 0.6 \times 0.4 = 2.88$

2. Calculate probabilities using the normal distribution: e.g. $X \sim N(30, 16)$.

   Find   **a)** $P(X < 20)$ **b)** $P(X < 35)$

   a) $P(X < 20) = P\left(Z < \dfrac{20-30}{4} = -2.5\right)$

   $\qquad\qquad = P(Z > 2.5) = 1 - 0.9938 = 0.0062$

   b) $P(X < 35) = P\left(Z < \dfrac{35-30}{4} = 1.25\right) = 0.8944$

Skills check:

1. If $X \sim B(8, 0.3)$, find the mean and variance of X.

2. $X \sim N(4, 9)$. Find

   a) $P(X < 10)$

   b) $P(X > 5)$

   c) $P(|X| < 3)$

9.1 Normal shape of some binomial distributions

Discrete random variables take only particular values, each with its own probability. Continuous random variables take values over an interval and probabilities are defined for ranges of values rather than individual values.

When there are a large number of possible values for a discrete distribution there can be a lot of calculations involved.

However, if we consider the graphs of some of the distributions, their shape looks similar to the normal distribution.

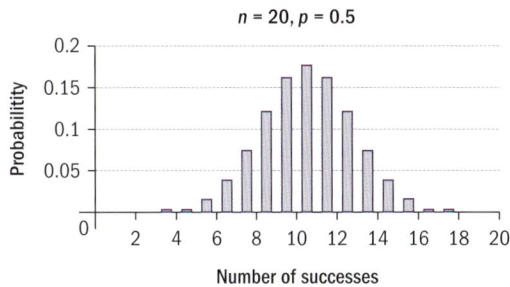

$n = 20, p = 0.5$

In section 7.3 we saw the probability distributions of $B(20, 0.25)$ and $B(20, 0.9)$ (reproduced below). $B(20, 0.25)$ looks approximately Normal but $B(20, 0.9)$ does not. Hence the shape being approximately Normal depends not only on n & p but also on n & q .

> We will deal more formally with this in section 9.3 but if $np > 5$, and $nq > 5$, then $B(n, p)$, will have an approximately Normal shape.

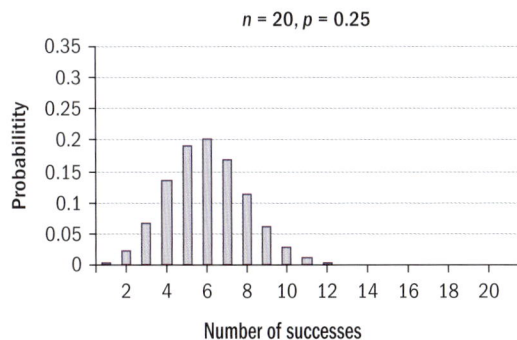

$n = 20, p = 0.25$

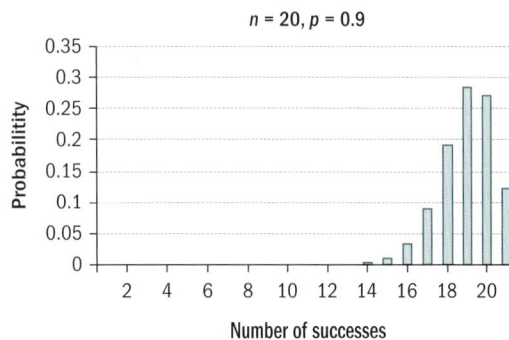

$n = 20, p = 0.9$

Exercise 9.1

1. Find the probabilities of all possible values for $B(6, 0.5)$ and represent these graphically as above (i.e. showing gaps between bars).

2. Find the probabilities of all possible values for $B(6, 0.1)$ and represent these graphically as above.

> Note that while even $B(6, 0.5)$ does not have the shape of the Normal properly defined yet, the shape of $B(6, 0.1)$ is completely unlike the Normal.

9.2 Continuity correction

In the diagrams in 9.1 the discrete probabilities are shown as bars with gaps between them. In fact, the bars really should have zero width since the probability occurs only at the integer value, in which case the diagrams would not look so much like the normal.

If we take '7' in the discrete distribution to be represented by the interval (6.5, 7.5) which are the values which round to 7, then the graphs look like:

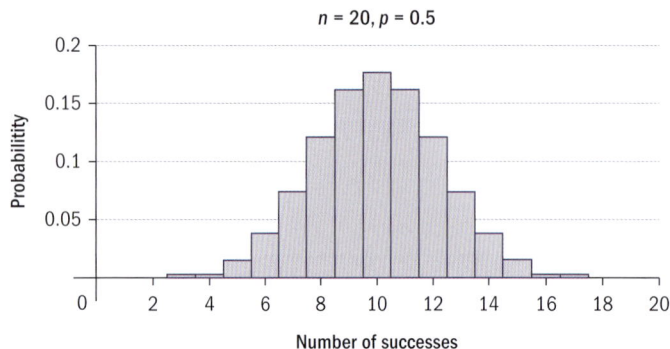

$n = 20, p = 0.5$

The resemblance to the normal distribution is now even stronger.

> When the normal is used to approximate the binomial (or any other distribution that takes only integer values) we must use a **continuity correction.**

If we want $\Pr\{X > 7\}$, we must use a cut off for the Normal at 7.5.
For $\Pr\{X < 7\}$ or $\Pr\{X \geq 7\}$ the cut off would be at 6.5.
The easiest way to determine whether it is at $+0.5$ or at -0.5 is to consider which two integers are to be separated.

Hence, $P(X = x) \overset{approx}{\approx} P(x - 0.5 < Y < x + 0.5)$,
where Y is the approximating normal random variable

Because the normal is continuous it doesn't matter whether you use $<$ or $\leq$ on either of these limits.

Example 1

Let $X \sim B(n, p)$ and $Y \sim N(np, npq)$ where n, p satisfy the conditions needed for Y to be used as an approximation for X and $q = 1 - p$.

Write down the probability you need to calculate for Y (including the continuity correction) as the approximation for each of the following probabilities for X.

a) $P(X < 15)$ **b)** $P(X > 12)$ **c)** $P(X \leq 17)$ **d)** $P(12 < X \leq 15)$

a) $P(Y < 14.5)$ ←   Separates up to 14 from 15 and up.

b) $P(Y > 12.5)$ ←   Separates up to 12 from 13 and up.

c) $P(Y \leq 17.5)$ ←   Separates up to 17 from 18 and up.

d) $P(12.5 \leq Y \leq 15.5)$ ←   The integers which satisfy this are 13, 14 and 15.

> Since the normal is continuous it requires only one calculation to do a probability for a block of individual values.

Exercise 9.2

1. Let $X \sim B(n, p)$ and $Y \sim N(np, np(1 - p))$ where n, p satisfy the conditions needed for Y to be used as an approximation for X. Write down the probability you need to calculate for Y (including the continuity correction) as the approximation for each of the following probabilities for X.

   a) $P(X < 42)$ **b)** $P(X > 31)$ **c)** $P(X \geq 9)$ **d)** $P(42 < X \leq 85)$.

2. Redraw as histograms, using the continuity correction, the probability diagrams you drew for Q1 and Q2 of exercise 9.1 (i.e. no gaps between bars).

9.3 The parameters for the normal approximation

In section 7.2 we saw that if $X \sim B(n, p)$, then $E(X) = np$ and $Var(X) = npq = np(1 - p)$.

If n is large and p is close to 0.5, so that the distribution is nearly symmetrical, then you can use the normal distribution to approximate the binomial.

The parameters to be used are the mean and variance of the binomial; $\mu = np$; $\sigma^2 = npq$.

The normal is symmetrical, and so is the binomial when $p = 0.5$, but as n gets larger the requirement for p to be close to 0.5 becomes less important.

> The general rule is that the normal can be used as an approximation when both np and nq are > 5.

Example 2

If $X \sim B(30, 0.4)$ calculate $P(12 \le X \le 15)$ by

a) calculating binomial probabilities   **b)** using a normal approximation.

a) $P(12 \le X \le 15) = P(X = 12, 13, 14, 15)$

$$= \binom{30}{12} 0.4^{12} 0.6^{18} + \binom{30}{13} 0.4^{13} 0.6^{17} + \binom{30}{14} 0.4^{14} 0.6^{16} + \binom{30}{15} 0.4^{15} 0.6^{15}$$

$= 0.9029$

$= 0.147375 + 0.136039 + 0.110127 + 0.078312$

$= 0.471853 = 0.472 \,(3 \text{ s.f.})$

b) For the binomial,

$\mu = np = 30 \times 0.4 = 12; \quad \sigma^2 = npq = 30 \times 0.4 \times 0.6 = 7.2,$

so use the $N(12, 7.2)$ distribution to approximate the $B(30, 0.4)$ distribution.

The continuity correction says

$P(12 \le X \le 15) \overset{approx}{\approx} P(11.5 < Y < 15.5)$ where Y is the approximating normal.

$P(11.5 < Y < 15.5) = P\left(\dfrac{11.5 - 12}{\sqrt{7.2}} < Z < \dfrac{15.5 - 12}{\sqrt{7.2}} \right)$

$= \Phi(1.304) - \Phi(-0.186) = 0.9032 - (1 - 0.5738) = 0.477 \,(3 \text{ s.f.})$

> You can see that the approximated calculation differs by less than 0.01 from the exact calculation. No matter how big the range of outcomes involved the normal approximation requires the same amount of work, whereas the binomial requires one calculation per outcome – here it was only 4 and therefore not too much effort was required, but to find $P(X \le 15)$ would require 16 separate binomial calculations.

Example 3

An airline estimates that 7% of passengers who book seats on a flight are 'no shows' – for one reason or another they miss the flight. On one flight, for which there are 185 available seats, the airline sells 197 tickets.

Find the probability that they will not have to refuse boarding to any passengers holding a valid ticket for that flight.

The number of 'no shows', X, can be modelled by a $B(197, 0.07)$ distribution if we make the simplifying assumption that all passengers behave independently of one another (almost certainly not entirely the case, but not wildly wrong in most cases – if there were a large group travelling together it would not be a good assumption).

We want $P(X \le 12)$.

Use $Y \sim N(197 \times 0.07, 197 \times 0.07 \times 0.93) = N(13.79, 12.8287)$ to approximate,

and calculate $P(Y < 11.5)$ using the continuity correction:

$P(Y < 11.5) = P\left(Z < \dfrac{11.5 - 13.79}{\sqrt{12.8287}} \right) = \Phi(-0.639) = 1 - 0.7386 = 0.2614 = 0.261 \,(3 \text{ s.f.}).$

Exercise 9.3

1. Which of the following could reasonably be approximated by a normal distribution?
   For those which can, give the normal which would be used.
   a) $X \sim B(50, 0.7)$ b) $X \sim B(10, 0.7)$ c) $X \sim B(500, 0.2)$ d) $X \sim B(500, 0.002)$

2. Use normal approximations to calculate
   a) $P(X < 42)$ if $X \sim B(50, 0.7)$ b) $P(X \geq 9)$ if $X \sim B(40, 0.3)$
   c) $P(X \geq 13)$ if $X \sim B(34, 0.37)$ d) $P(25 < X \leq 37)$ if $X \sim B(80, 0.4)$
   e) $P(43 \leq X \leq 55)$ if $X \sim B(124, 0.43)$.

3. For $X \sim B(50, 0.3)$
   a) calculate $P(12 < X < 20)$
      i) using tables   ii) using a normal approximation.
   b) i) What error is there in using the normal approximation?
      ii) Express part (b) (i) as a percentage of the exact probability.

4. Explain briefly the conditions under which the normal distribution can be used as an approximation to a binomial distribution. If the conditions are satisfied, state what normal distribution would be used to approximate the $X \sim B(n, p)$ distribution.

5. On a production line, on average 6% of the bottles of lemonade are not filled properly.
   a) If five bottles are examined, find the probability that exactly one of them is not filled properly.
   b) If 2000 bottles are examined, use a normal approximation to find the probability that less than 100 of them are not filled properly.

6. A multiple choice test has 50 questions each with four possible answers.
   a) If Catharine guesses the answer to each question randomly, state the exact distribution of X, the number of answers Catharine gets correct.
   b) If the test has a pass mark of 20, find the probability that Catharine passes the test.
   c) Shopna takes the same test, but knows enough to be able to rule out one of the possible answers to each question. Use a normal approximation to find the probability that Shopna fails the test.

7. A Health Trust gives guidance to doctors that they should show any patients with raised blood pressure some information on lifestyle which would help bring down their blood pressure.
   The doctors in one surgery ask their research assistant to look at the patient records and find that 35% of their patients have raised blood pressure.
   a) The surgery has 84 patients booked for appointments on the following day. Assuming that these 84 patients are a random sample from all the patients registered with that surgery, find the probability that at least 25 will have raised blood pressure.
   b) Comment on the assumption made in part (a) that the people with appointments at the surgery on the following day will be a random sample from all people on the register.

Summary exercise 9

1. Amil is a dentist who finds that, on average, one in five of his patients do not turn up for their appointment.

   a) using a binomial distribution find the mean and variance of the number of patients who turn up for their appointment in a clinic where there were 20 appointments.

   b) find the probability that more than 2 patients do not turn up for their appointment in the clinic in part a.

   c) using a suitable approximation find the probability he sees more than 135 patients in a week where he has 154 appointments.

2. A fair coin is tossed repeatedly. Using suitable approximations where appropriate, find the probability that you would see

   a) more than 7 heads in 10 tosses

   b) more than 70 heads in 100 tosses

   c) more than 700 heads in 1000 tosses

3. A fair dice is thrown repeatedly. Using suitable approximations where appropriate, find the probability that you would see

   a) more than 3 sixes in 12 throws

   b) more than 30 sixes in 120 throws

   c) more than 300 sixes in 1200 throws

EXAM-STYLE QUESTIONS

4. A golfer practises on a driving range. His objective is to hit a ball to within 10 m of a flag.

   a) On his first visit the probability of success with each particular ball is 0.3. If he hits ten balls what is the probability of

      i   four or fewer successes

      ii four or more successes?

   b) Using a suitable approximation, some weeks later the probability of success has increased to 0.53. What is the probability of 120 or more successes in 250 drives?

   c) A year later the probability of success has increased to 0.92. What is the probability of 3 or fewer failures in 50 drives?

5. Among the blood cells of a certain animal species, the proportion of cells which are of type O is $\frac{1}{3}$ and the proportion of cells which are of type AB is 0.005.

   a) Find the probability that in a random sample of 8 blood cells at least 2 will be of type O.

   Using suitable approximations, find the probability that

   b) in a random sample of 200 blood cells the total number of type O and type AB cells is at least 81

6. A tour operator organises a visit for cricket enthusiasts to India in November.

   The package includes a ticket for a one-day international in Nagpur. Places on the tour must be booked three months in advance. From past experience, the tour operator knows that the probability of a person who has booked a place subsequently withdrawing is 0.1 and is independent of other withdrawals.

   a) Twenty-five people book places. Find the probability that

      i)   none withdraw,

      ii) two or more withdraw.

      iii) The tour operator only has 21 tickets available for the one day international. What is the probability that he will be able to provide everyone who goes on tour with a ticket?

b) An organiser of a similar but larger tour accepts 250 bookings but has only 210 tickets for the one-day international. Find, using a suitable approximation, the probability that this organiser will be able to provide everyone on this tour with a ticket. (Assume that the probability of a person withdrawing remains at 0.1)

7. On a certain road 20% of the vehicles are trucks, 16% are buses and the remainder are cars.

   i) A random sample of 11 vehicles is taken. Find the probability that fewer than 3 are buses. [3]

   ii) A random sample of 125 vehicles is now taken. Using a suitable approximation, find the probability that more than 73 are cars. [5]

Cambridge International AS and A Level Mathematics 9709, Paper 6 Q3 May/June 2009

8. A die is biased so that the probability of throwing a 5 is 0.75 and the probabilities of throwing a 1, 2, 3, 4 or 6 are all equal.

   i) The die is thrown three times. Find the probability that the result is a 1 followed by a 5 followed by any even number. [3]

   ii) Find the probability that, out of 10 throws of this die, at least 8 throws result in a 5. [3]

   iii) The die is thrown 90 times. Using an appropriate approximation, find the probability that a 5 is thrown more than 60 times. [5]

Cambridge International AS and A Level Mathematics 9709, Paper 6 Q7 May/June 2008

9. When a butternut squash seed is sown the probability that it will germinate is 0.86, independently of any other seeds. A market gardener sows 250 of these seeds. Use a suitable approximation to find the probability that more than 210 germinate. [5]

Cambridge International AS and A Level Mathematics 9709, Paper 61 Q1 October/ November 2011

10. Kamal has 30 hens. The probability that any hen lays an egg on any day is 0.7. Hens do not lay more than one egg per day, and the days on which a hen lays an egg are independent.

    i) Calculate the probability that, on any particular day, Kamal's hens lay exactly 24 eggs. [2]

    ii) Use a suitable approximation to calculate the probability that Kamal's hens lay fewer than 20 eggs on any particular day. [5]

Cambridge International AS and A Level Mathematics 9709, Paper 6 Q4 May/June 2003

11. It is known that, on average, 2 people in 5 in a certain country are overweight. A random sample of 400 people is chosen. Using a suitable approximation, find the probaility that fewer than 165 people in the sample are overweight. [5]

Cambridge International AS and A Level Mathematics 9709, Paper 6 Q1 May/June 2005

12. In tests on a new type of light bulb it was found that the time they lasted followed a normal distribution with standard deviation 40.6 hours. 10% lasted longer than 5130 hours.
   i)   Find the mean lifetime, giving your answer to the nearest hour. [3]
   ii) Find the probability that a light bulb fails to last for 5000 hours. [3]
   iii) A hospital buys 600 of these light bulbs. Using a suitable approximation, find the probability that fewer than 65 light bulbs will last longer than 5130 hours. [4]
   Cambridge International AS and A Level Mathematics 9709, Paper 6 Q7 October/November 2005

13. A shop sells old video tapes, of which 1 in 5 on average are known to be damaged.
   i)   A random sample of 15 tapes is taken. Find the probability that at most 2 are damaged. [3]
   ii) Find the smallest value of n if there is a probability of a least 0.85 that a random sample of n tapes contains at least one damaged tape. [3]
   iii) A random sample of 1600 tapes is taken. Use a suitable approximation to find the probability that there are at least 290 damaged tapes. [5]
   Cambridge International AS and A Level Mathematics 9709, Paper 6 Q7 May/June 2004

14. Assume that, for a randomly chosen person, their next birthday is equally likely to occur on any day of the week, independently of any other person's birthday. Find the probability that, out of 350 randomly chosen people, at least 47 will have their next birthday on a Monday. [5]
   Cambridge International AS and A Level Mathematics 9709, Paper 61 Q2, May/June 2013

Chapter summary
- The binomial distribution $B(n, p)$ can be approximated by the normal distribution $N(np, npq)$ provided both np and $nq = n(1-p)$ are greater than 5.
- When the normal is used to approximate the binomial (or any other distribution that takes only integer values), a continuity correction must be used.

Review exercise C

1. If $X \sim B(8, 0.4)$ find the probability that
   i) $X = 2$ ii) $X = 5$

2. If $X \sim B(10, 0.12)$ find
   i) $P(X > 2)$ ii) $P(X \leq 8)$

3. X is a Binomial random variable based on 20 trials. $E(X) = 17$.
   a) Find the standard deviation of X.
   b) Find $P(X > \mu)$, where $\mu = E(X)$.

4. A quality control agent tests sets of 5 components from a production line which is known to produce 95% good components.
   a) Find the probability that a set chosen at random is free from defects.
   b) If the quality control agent tests five sets before lunch, find the probability that four of these sets were free from defects.
   c) The agent tests 120 sets in a week's work. On average, how many sets does she find which are not perfect?

5. The table below shows, for a particular population, the proportions of people in each of the four main blood groups.

Blood group	O	A	B	AB
Proportion	0.40	0.28	0.20	0.12

   A random sample of 500 people is selected.
   Using suitable approximations estimate the probability that the sample contains:
   a) at most 90 people with blood group B
   b) at least 55 people with blood group AB
   c) more than 200 but fewer than 250 people with blood group O.

6. The random variable is normally distributed with mean 152 and variance 144.
   Find
   a) $P(X < 176)$,   b) $P(140 < X < 170)$.
   It is known that $P(a \leq X) = 0.15$.
   c) Find the value of a.

7. Bottles of orange juice have a stated volume of 500 ml. The bottles are filled by a machine, which dispenses a volumes which is normally distributed with mean 505 ml and standard deviation 3 ml.
   a) Find the probability of a bottle containing less than the stated volume.

   The bottles have a capacity of 510 ml. Any time that the machine dispenses more than this, the extra just spills out.
   b) Find the probability of a bottle containing less than the stated volume, given that it didn't overflow.

   A new machine is installed. Only 0.5% of bottles overflow when the mean amount dispensed by the new machine is set to 505 ml.
   c) Find the standard deviation of the amount dispensed by the new machine, correct to 1 decimal place.

8. A machine is set to produce bolts with a mean length of 35 mm. 7% of the bolts have to be rejected because they do not meet the minimum length of 32 mm.
   a) Find an estimate of the standard deviation of the lengths of the bolts.
   b) The bolts cannot be longer than 40 mm. Calculate the proportion of bolts that are rejected as too long.
   c) Find the probability that a bolt which is known not to be too short is rejected because it is too long.

9. A machine is known to produce pencils with a standard deviation in length of 5 mm, and the mean length can be set to whatever is wanted.

   A client specifies that a maximum of 3% of pencils are to be less than 20 cm long. What should the machine operator set the mean at?

10. A medical researcher has a random sample of the lengths of 500 index fingers from a particular tribe of people. She finds that 73 of them are less than 7 cm long and 145 are longer than 10 cm.
Find estimates of the mean and standard deviation of the lengths of index fingers in that tribe.

11. A multiple choice test has 40 questions with 5 possible answers of which only 1 is correct. For each question on the test the correct answer scores 4 and any incorrect answer (a distractor) scores –1.
    a) Show that if a person guesses randomly that the mean score on the test will be zero.

    Katerina finds that she can discount two of the distractors on three quarters of questions on the test but the others she would have to guess. She does not attempt the questions which would be a complete guess but does guess on the questions she can rule out two of the distractors.
    b) Find the mean and variance of her score on the test.
    c) Use a suitable approximation to estimate the probability she gets at one third of all the questions correct on the test.

12. i) A manufacturer of biscuits produces 3 times as many cream ones as chocolate ones. Biscuits are chosen randomly and packed into boxes of 10. Find the probability that a box contains equal numbers of cream biscuits and chocolate biscuits. [2]
    ii) A random sample of 8 boxes is taken. Find the probability that exactly 1 of them contains equal numbers of cream biscuits and chocolate biscuits. [2]
    iii) A large box of randomly chosen biscuits contains 120 biscuits. Using a suitable approximation, find the probability

that it contains fewer than 35 chocolate biscuits. [5]

Cambridge International AS and A Level Mathematics 9709, Paper 6 Q6, October/November 2002

13. The distance in metres that a ball can be thrown by pupils at a particular school follows a normal distribution with mean 35.0 m and standard deviation 11.6 m.
    i) Find the probability that a randomly chosen pupil can throw a ball between 30 and 40 m. [3]
    ii) The school gives a certificate to the 10% of pupils who throw further than a certain distance.

    Find the least distance that must be thrown to qualify for a certificate. [3]

Cambridge International AS and A Level Mathematics 9709, Paper 6 Q3, October/November 2002

14. The random variable X is the length of time in minutes that Jannon takes to mend a bicycle puncture. X has a normal distribution with mean μ and variance σ^2. It is given that $P(X > 30.0) = 0.1480$ and $P(X > 20.9) = 0.6228$. Find μ and σ. [5]

Cambridge International AS and A Level Mathematics 9709, Paper 61 Q3, May/June 2010

15. i) The daily minimum temperature in degrees Celsius (°C) in January in Ottawa in a random variable with distribution N(–15.1, 62.0). Find the probability that a randomly chosen day in January in Ottawa has a minimum temperatue above 0°C. [3]
    ii) In another city daily minimum temperature in °C in January is a random variable with distribution N(μ, 40.0). In this city the probability that a randomly chosen day in January has a minimum temperature above 0°C is 0.8888. Find the value of μ. [3]

Cambridge International AS and A Level Mathematics 9709, Paper 6 Q3, October/November 2008

16. a) The random variable X is normally distributed. The mean is twice the standard deviation. It is given that $P(X > 5.2) = 0.9$. Find the standard deviation. [4]

b) A normal distribution has mean μ and standard deviation σ. If 800 observations are taken from this distribution, how many would you expect to be between $\mu - \sigma$ and $\mu + \sigma$? [3]

Cambridge International AS and A Level Mathematics 9709, Paper 6 Q3, May/June 2007

17. In the holidays Martin spends 25% of the day playing computer games. Martin's friend phones him once a day at a randomly chosen time.

i) Find the probability that, in one holiday period of 8 days, there are exactly 2 days on which Martin is playing computer games when his frind phones. [2]

ii) Another holiday period lasts for 12 days. State with a reason whether it is appropriate to use a normal approximation to find the probability that there are fewer than 7 days on which Martin is playing computer games when his friend phones. [2]

iii) Find the probability that there are at least 13 days of a 40-day holiday period on which Martin is playing computer games when his friend phones. [5]

Cambridge International AS and A Level Mathematics 9709, Paper 61 Q5, May/June 2010

18. In a normal distribution, 69% of the distribution is less than 28 and 90% is less than 35. Find the mean and standard deviation of the distribution. [6]

Cambridge International AS and A Level Mathematics 9709, Paper 6 Q3, October/November 2003

19. Single cards, chosen at random, are given away with bars of chocolate. Each card shows a picture of one of 20 different football players. Richard needs just one picture to complete his collection. He buys 5 bars of chocolate and looks at all the pictures. Find the probability that

i) Richard does not complete his collection. [2]

ii) he has the required picture exactly once, [2]

iii) he completes his collection with the third picture he looks at. [2]

Cambridge International AS and A Level Mathematics 9709, Paper 6 Q4, October/November 2003

Maths in real-life

Statistics is definitely not a lonely world

Most statisticians work in inter-disciplinary teams – they may work on environmental issues, in business and finance, in manufacturing, in health, investigating effective ways of tackling poverty or almost any interesting human activity – working in conjunction with specialists in the area.

One of the career growth areas in recent years is simulation modeling. Computers now offer such immense power that complex simulations can be run on even standard desktop or laptop computers. Many interesting situations like the spread of disease and financial investment risk analysis can be modeled by relatively simple steps repeated many times. With the advent of air travel and the ability for disease to spread rapidly, these computer models are extremely important.

Many diseases follow the basic SIR model – the population is made up of three groups: *Susceptible, Infected,* and *Recovered* (the recovered group can be thought of as including anyone who is immune for any other reason, such as vaccination). Each disease will have varying characteristics – how long someone is infected for, how likely they are to pass it on to a person they come in contact with etc. A simulation can build these in as parameters, so for a disease where a person is infectious for 3 days, and the probability of infecting a person on any given day is 0.2, the chance of not infecting them in 3 days will be $0.8^3 = 0.512$ or roughly evens. However, If you want to model the spread in a population you have to take account of multiple interactions over the period of time the disease persists.

Setting up a computer model of the disease behavior and running it repeatedly allows health professionals to explore different possible courses of action they may take to intervene in trying to control epidemics.

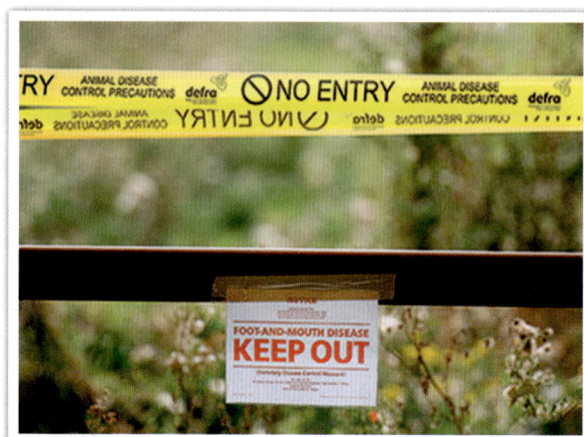

Being able to model the patterns of disease is extremely important. For example, Foot and Mouth disease is a highly infectious viral disease affecting cloven-hoofed animals which caused an epidemic in the UK in 2001. The disease is extremely hard to contain and there were 2000 cases in farms across the country. Some infected animals had been transported to the continent before the outbreak was identified and measures were taken to prevent the epidemic becoming continent wide – for example in Holland they slaughtered a quarter of a million cattle as a precaution.

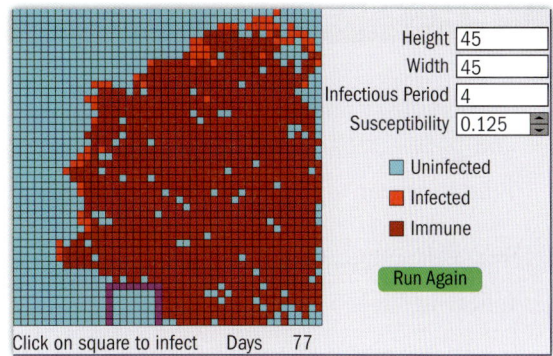

The economic implications of finding effective ways of containing virulent diseases like this are huge. How would a government manage if there was an outbreak of a disease with these characteristics amongst humans?

A number of simulations modeling the spread of disease can be found on the website, http://www.oxfordsecondary.com/cambridge-alevelmath. These will give you some insights into how a model is developed – introducing greater complexity at each stage, so it starts with a static model, where infected people have fixed contact points with other people.

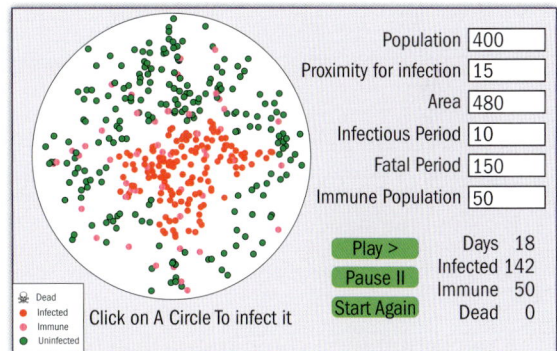

The first screen shot on the right shows the model where the user can select 'immune' squares (shown in purple at the centre bottom of the grid), which can be thought of as representing barriers – such as a river, or a mountain range – that the disease cannot pass over but has to move around. In this instance almost the entire population was infected, with only a small number (the light blue cells) having escaped by chance, except for the block shielded by the barrier.

The second screen shot shows the final model in the series. The population is moving around and there are more parameters to control – allowing you to explore what happens with different combinations of effects. 'People' who get infected in this model either die (shown as skull) or recover (shown as purple circle).

Paper A

1. The lengths, x cm, of 25 rulers wee measured, and are summarised by $\sum(x - 30) = 0.6$ and $\sum(x - 30)^2 = 5.32$

   Find the mean and standard deviation of the rulers. [3]

2. The distances run in training by a group of marathon runners is summarised in the following table.

Distance run (x km)	Frequency
$15 \leq x < 25$	19
$25 \leq x < 30$	17
$30 \leq x < 35$	24
$35 \leq x < 40$	12

   Draw a histogram on graph paper to represent this data. [4]

3. Lengths of climbing rope are Normally distributed with mean 24.2 metres, and 20% of the ropes are longer than 25 metres.

   i) Find the standard deviation of the lengths of the ropes. [3]

   ii) Find the probability that a randomly selected rope is shorter than 24 metres. [3]

4. A nurse running a clinic knows that, on average, 22% of her appointments do not turn up.

   i) Find the probability that, in a clinic with 10 appointments, fewer than 7 patients turn up. [4]

   ii) For five randomly selected clinics with 10 appointments, find the probability she has at least one clinic with fewer than 7 patients. [3]

5. A doctor specializes in sports injuries. The 134 injuries he treats in some sports over a month are summarised in the following table.

	knee	shoulder	other	
Basketball	23	10	7	40
Tennis	12	14	16	42
Soccer	22	6	24	54
	57	30	47	

   A patient is chosen at random from this group. What is the probability that the patient

   i) suffered a knee injury playing tennis [2]

   ii) suffered a shoulder injury [2]

   iii) was playing soccer, given that they suffered an injury classified as other. [3]

6. **a)** A group of 12 people consist of 3 boys, 5 girls and 4 adults. In how many ways can a team of 4 be chosen if

      i) at least two boys are in the team [2]

      ii) the adults are either all in the team or all not in the team [2]

      iii) there must be at least one of each of the boys, girls and adults [3]

  b) Find the number of distinct permutations of the letters of the word ASSESS [3]

7. A spinner has 5 equal sections numbered 1, 2, 3, 4, 5.

  i) Find the probability of obtaining at least 6 odd numbers in 8 spins. [4]

The spinner is spun twice. Let X be the product of the two scores. The following table shows the possible values of X.

X	1	2	3	4	5
1	1	2	3	4	5
2	2	4	6	8	10
3	3	6	9	12	15
4	4	8	12	16	20
5	5	10	15	20	25

  ii) Draw up a table showing the probability distribution of X. [3]

  iii) Calculate the mean and variance of X. [4]

  iv) Find the probability that a randomly chose observation of X is greater than the mean of X. [2]

1. It is given that $X \sim N(46.3, 12.3)$. Find the probability that a randomly chosen value of x lies between 40 and 50. [3]

2. The following back-to-back stem-and-leaf diagram show the pulse rates of a group of students before and after a short period of moderate exercise.

Before exercise						stem	After exercise							
(3)				9	7	3	**6**								
(6)	9	8	6	3	1	0	**7**	9							(1)
(5)		8	6	4	2	1	**8**	2	2	3	4	5	7	9	(7)
(1)						2	**9**	2	5	6	8	9			(5)
							10	2	5						(2)

Key 2 | 8 | 3 means 82 beats per minute before exercise and 83 beats per minute after exercise

   i) Find the median and quartiles of the pulse rates before exercise. [2]

   You are given that the median pulse rate after exercise is 89, the lower quartile is 83 and the upper quartile is 98.

   ii) Represent the data by means of a pair of box-and-whisker plots in a single diagram. [3]

3. The following table shows the results of a survey to find the average daily time, in minutes, that exercising on the previous day.

Time (t minutes)	Frequency
$0 \le t < 10$	3
$10 \le t < 20$	5
$20 \le t < 40$	f
$40 \le t < 70$	4
$70 \le t < 120$	3

The mean time was calculated to be 35.8 minutes.

   i) Form an equation involving f and hence show that the total number of children was 25. [4]

   ii) Find the standard deviation of the times. [2]

4. It is known that, on average, 15% of people in a certain country suffer from malnutrition.

   a) Find the probability that in a random sample of 10 people there is not more than one person suffering from malnutrition. [3]

   b) Another randomly selected sample of 400 people is now chosen. Find the probability that not more than 40 people are suffering from malnutrition. [4]

5. A player has a probability of 0.4 of winning a tennis tournament, and a probability of 0.2 of not reaching the semi-final stage. If he wins the tournament, the corresponding probabilities for the next tournament are 0.5 and 0.1. If he reaches the semi-final but does not win the first tournament the corresponding probabilities are 0.5 and 0.3.
If he fails to reach the semi-final the corresponding probabilities are 0.2 and 0.6.

   i) Find the probability that he does not reach the semi-final stage in the first tournament and wins the second tournament. [2]

   ii) Find the probability that he wins at least one tournament. [3]

   iii) Given that he wins the second tournament, what is the probability that he also won the first tournament. [3]

6. The random variable X has probability function $P(X = x) = kx^2 \quad x = 1, 2, 3, 4, 5$

   i) Show that $k = \dfrac{1}{55}$ [2]

   ii) Find the probability $P(X \leq 3)$ [2]

   iii) Calculate $E(X)$ [3]

   iv) Find the probability that four randomly chosen observations of x contain exactly one which is greater than $E(X)$. [4]

7. a) Four friends attend different schools, and each brings two classmates to a quiz evening. The quiz is for teams of 6, so the group make up two teams.

   i) How many different pairs of teams are possible if students at the same school are always on the same team. [2]

   ii) How many different pairs of teams are possible if the four friends are all on one team. [2]

   iii) How many different pairs of teams are possible if each team includes at least one student from each school. [3]

   b) Find the number of distinct permutations of the letters of the word DIFFERENCE. [3]

Answers

The answers given here are concise. However, when answering exam-style questions, you should show as many steps in your working as possible.

Cambridge International Examinations bears no responsibility for the example answers to questions taken from its past paper questions which are contained in this publication.

1 Introduction to statistical thinking

Exercise 1.1 page 4

1. Long-term form plays an important part in an extended league format. However, a "surprise" result in a knock-out tournament can be caused by a temporary lack of form or from a team playing above their normal standard.

2. Either collect or use previously recorded data to estimate the travel times for the journeys involved.

3. There is extensive research carried out by bodies like the World Health Organisation (WHO) into this question. Unfortunately much of this information is often inconclusive and the outcomes can also be confused by reports from interested parties such as the mobile phone providers.

Exercise 1.3 page 11

1.

Days late	0	1	2	3	4	5
Frequency	19	7	2	1	0	1

2.

Depth (feet)	3.6 – 4	4.1 – 4.5	4.6 – 5	5.1 – 5.5	5.6 – 6	6.1 – 6.5
Frequency	3	4	4	4	8	6

3. **a)** Make of car (qualitative, observed)
Colour of car (qualitative, observed)
Year of registration (quantitative, discrete, may be determined from the number plate)
Mileage (quantitative, continuous, read from milometer)

b) Length of race (quantitative, continuous, measured or stated in race information)
Number of entries (quantitative, discrete, counted or determined from the entry list)
Winning time (quantitative, continuous, measured)
Make of winning bike (qualitative, observed)

c) Number of bulbs (quantitative, discrete, counted)
Size (quantitative, continuous, measured)
Colour (qualitative, observed)
Make (qualitative, observed)

d) Name (qualitative, observed)
Location of source (qualitative, observed)
Length (quantitative, continuous, measured)
Rate of flow at various points (quantitative, continuous, measured)

e) Number of sheep (quantitative, discrete, counted)
Colour (qualitative, observed)
Place of origin (qualitative, observed)
Weight (quantitative, continuous, measured)

f) Number produced per day (quantitative, discrete, counted)
Type of chocolate (qualitative, observed)
Name (qualitative, observed)
Weight (quantitative, continuous, measured)

g) Consistency (e.g. slushy) (qualitative, observed)
Air temperature (quantitative, continuous, measured)
Depth (quantitative, continuous, measured)
Duration of snowstorm (quantitative, continuous, measured)

h) Name (qualitative, observed)
Height (quantitative, continuous, measured)
Team (qualitative, observed)
Number of runs scored (quantitative, discrete, counted)

i) Gender (qualitative, observed)
Age (quantitative, continuous, measured)
Height (quantitative, continuous, measured)
Loudness of roar in decibels (quantitative, continuous, measured)

j) Year of first opening to the public (quantitative, discrete, research historical records)
Number of visitors per day (quantitative, discrete, counted)
Age (quantitative, continuous, measured)
Height (quantitative, continuous, measured)

k) Types of business (qualitative, observed)
Number of business (quantitative, discrete, counted)
Height of tallest building (quantitative, continuous, measured)
Annual turnover (quantitative, discrete, counted)

l) Type of vehicles (qualitative, observed)
Number of vehicles (quantitative, discrete, counted)
Number of traffic police (quantitative, discrete, counted)
Time taken to travel one mile (quantitative, continuous, measured)

2 Measures of location and spread

Skills check page 14

1. a) 2   b) 2   c) 5   d) 2
2. a) 6.8 (1 d.p.)   b) 6.5   c) 2
   d) 6

Exercise 2.1 page 17

1. a) 77   b) 76.5   c) 77
2. a) 4   b) 4   c) 4.1 (1 d.p.)
3. 90.4 beats per minute (1 d.p.), 109.1 beats per minute (1 d.p.)
4. 65.0 cm (1 d.p.)
5. 21.1 minutes
6. $16 600

Exercise 2.2 page 20

1. $Q_1 = 75$, $Q_3 = 77$
2. $Q_1 = 3$, $Q_3 = 5$
3. a) 64 beats per minute   b) 8.5 beats per minute
4. a) 28   b) 11
5. a) 49.5 seconds   b) 58 seconds
6. a) 19.5 minutes   b) 18 minutes

Exercise 2.3 page 25

1. a) mean = 13.8 (3 s.f.); standard deviation = 3.86 (3 s.f.)
   b) mean = 6.38 (3 s.f.); standard deviation = 1.37 (3 s.f.)
2. mean = 74.3 cm (3 s.f.); standard deviation = 1.99 cm (3 s.f.)
3. mean = 346 g (3 s.f.); standard deviation = 11.5 g (3 s.f.)
4. mean = 2.5; variance = 3.52 (3 s.f.)
5. mean = 0.3; standard deviation = 3.43 (3 s.f.)
6. 2.77 (3 s.f.)
7. 23.1 (3 s.f.)
8. Standard deviation is approximately 10.

Exercise 2.4 page 29

1. mean = 80.5, variance = 132.5
2. mean = 1137, variance = 322

3. a) 55.85 °F   b) 10, 12, 11.5, 13, 13, 14, 15.5, 17
   c) 13.25 °C
4. a) mean = 11.2 (3 s.f.), variance = 2.64 (3 s.f.)
   b) $X = 0, 1, 2, 3$
   c) mean = 1.08 (3 s.f.), variance = 0.660 (3 s.f.)
5. a) $X = 0, 3, 5, 7, 9, 14$
   b) mean = 4.46 (3 s.f.), variance = 7.48 (3 s.f.)
   c) mean = 16100 (3 s.f.), variance = 46700000 (3 s.f.)
6. a) $T = 0, 1, 2, 3$
   b) mean = 1.15 (3 s.f.), variance = 0.873 (3 s.f.)
   c) mean = 18.2 (3 s.f.), variance = 21.8 (3 s.f.)

Summary exercise 2 page 31

1. 6; 25.4
2. −1.85; 8.55
3. 799; 7188
4. a) mean = 65.6 (3 s.f.), variance = 31.9 (3 s.f.)
   b) 61523.646   c) 64.7 (3 s.f.)
5. median = 8, interquartile range = 9.5
6. a) mean = 67.6, standard deviation = 6.25 (3 s.f.)
   b) $a = 36.35$, $b = 0.625$
7. i) 12   ii) 8.88 (3 s.f.)
8. Median = $47 000.
9. i) Proof   ii) 16.1 minutes (3 s.f.)
10. mean = 38.4 mm, standard deviation = 4.57 mm (3 s.f.)
11. $\Sigma x = 804$, $\Sigma x^2 = 27011.76$
12. i) 2.1   ii) $c = 78.2$
13. 26; 257
14. 33.75 minutes; 2.3 minutes

3 Representing and analysing data

Skills check page 34

1. a) median = 12   lower quartile = 9   upper quartile = 19
   b) median = 4   lower quartile = 4   upper quartile = 5
   c) median = 26.1 (1 d.p.)   lower quartile = 17.1   upper quartile = 40.2 (1 d.p.)

Exercise 3.1 page 38

1. **a)** Type A: median = 44 g; inter-quartile range = 8 g
   Type B: median = 51 g; inter-quartile range = 6 g

   b)

Type A		Type B
5	2	
4 2 0	3	
9 7 7 5 5	3	
4 4 4 4 2 2 2 2 1	4	4
8 7 7 7 7 6 6 6 5 5	4	5 6 7 7 8 8 8 9 9 9 9
2 2 1 0	5	0 0 0 0 1 1 1 1 1 2 2 3 3 4
9 8 7	5	5 5 6 6 7 7 8 9 9

   Key: 5|4|6 means 45 g for Type A and 46 g for Type B

   c) On average Type B plums are heavier than Type A plums. Type A plums are more variable in weight than Type B plums.

   d) Type B because they are heavier and less variable.

2. **Before exercise:** median = 81 beats/minute, inter-quartile range = 20 beats/minute
   After exercise: median = 94 beats/minute, inter-quartile range = 13 beats/minute
   Pulse rates were higher on average and less variable after exercise than before.

3. **January:** median = 22.55 g, inter-quartile range = 2.25 g
   April: median = 21.5 g, inter-quartile range = 0.7 g
   On average the Dunnocks are heavier and their weights are more variable in January than in April.

4. **Males:** median = 195.5 mm, inter-quartile range = 4 mm
   Females: median = 232.5 mm, inter-quartile range = 8 mm
   On average the females have a longer wingspan than the males and females have a more variable wingspan than the males.

5.

Fruit		Vegetables
3	0	2 2 2 3 4 4 5 5 5 6 7 8
9 9 5 3 2 2 1	1	1 8
6 6 3 1 0	2	6
4 0	3	

   Key: 3|2|6 means 23 g for Fruit and 26 g for Vegetables
   Fruit: median = 9 g, inter-quartile range = 14 g
   Vegetables: median = 5 g, inter-quartile range = 5 g
   The amount of carbohydrate in fruit is larger on average and more variable than in vegetables.

Exercise 3.2 page 41

1. On average resort B is hotter than resort A. Temperatures are more variable in resort A than in resort B. It would also be useful to have information about the temperature at different times of the year, hours of sunshine and the amount of rainfall.

2. **a)** On average males are paid more than females. The overall range is similar for males and females but the very low paid males are outliers. The inter-quartile range is less for males so their salaries are less variable than the salaries of females.

   b) The average starting salary is similar for males and females with the females' salaries varying more than the males'. It may be that the males stay with the college longer which could account for their average salaries being higher in 1991–1992.

3. **a)**

   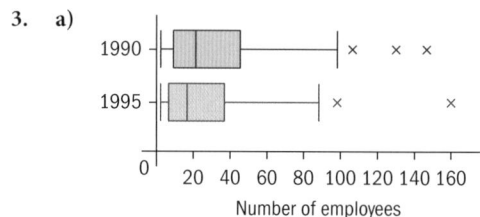

   b) The number of employees is higher on average and more variable in 1990 than in 1995.

4. **a)**

	Min	LQ	Median	UQ	Max
Fruit	30	126	147	161	280
Vegetables	78	85	99	148	148
Seafood	84	84	84	84	84

   b) The 5 summary statistics are equal.

   c)

   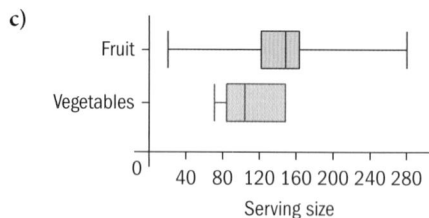

5. On average the highest scores were in the long jump and the lowest scores were in the 1500 metres. The overall range was highest for the 1500 metres and lowest for the 100 metres. The inter-quartile range was highest for the long jump and lowest for the 100 metres. There was less variation in the scores for the 100 metres than the other two events.

Exercise 3.3 page 44

1.

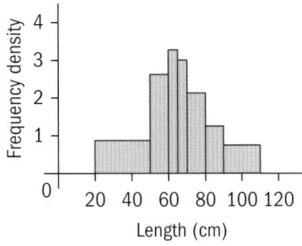
Frequency density vs Length (cm)

2. a)

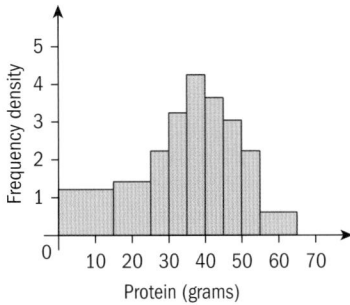
Frequency density vs Protein (grams)

b) mean = 34.2 g (1 d.p.),
standard deviation = 14.7 g (1 d.p.)

c) On average the daily protein intake was higher
in the second country than in the first. The daily
protein intake was more variable in the first
country than in the second.

3.

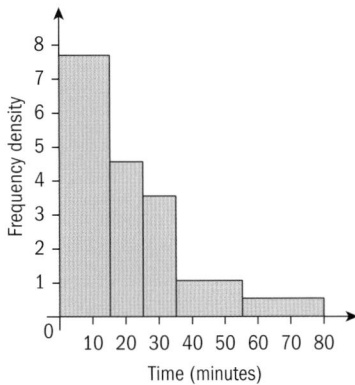
Frequency density vs Time (minutes)

4.

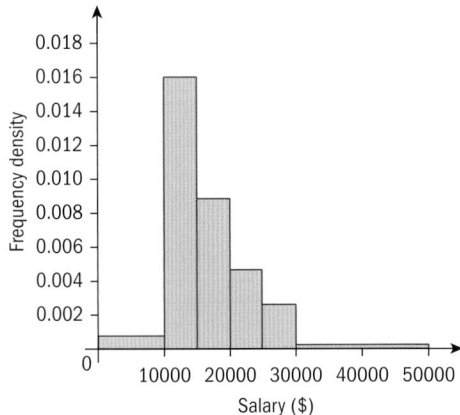
Frequency density vs Salary ($)

5. a)

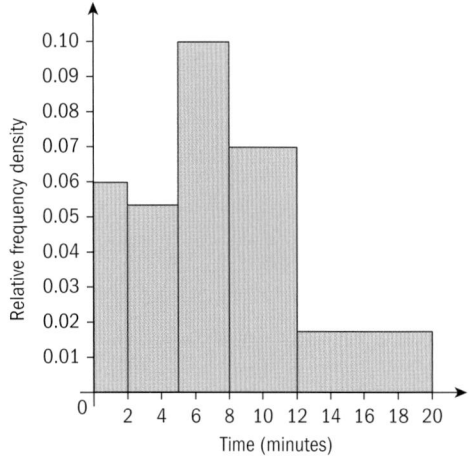
Relative frequency density vs Time (minutes)

b) 10

Exercise 3.4 page 48

1. a) median = 64, lower quartile = 60.5,
upper quartile = 69

b)

Pulse rate	Frequency
51–55	4
56–60	5
61–65	12
66–70	8
71–75	4
76–80	2
81–85	1

c) median = 64.5 (1 d.p.), lower quartile = 60.7
(1 d.p.), upper quartile = 69.6 (1 d.p.)

2. a) median = 28, lower quartile = 23,
upper quartile = 34

b)

Age	Frequency
17–20	7
21–24	6
25–28	12
29–32	7
33–48	13

c)

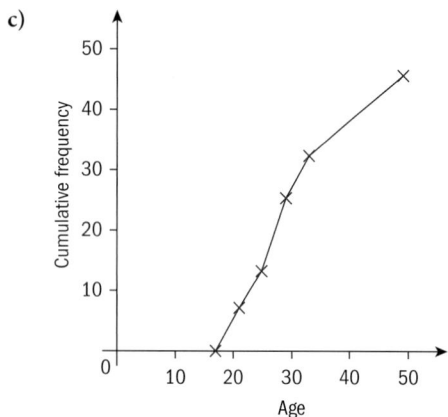

d) median = 28, lower quartile = 24,
upper quartile = 35

3. **a)**

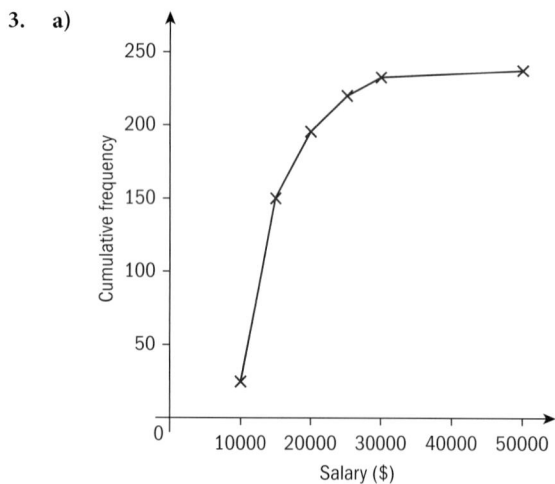

b) median = $13 700, lower quartile = $11 400,
upper quartile = $18 000

c) 4%

4. **a)**

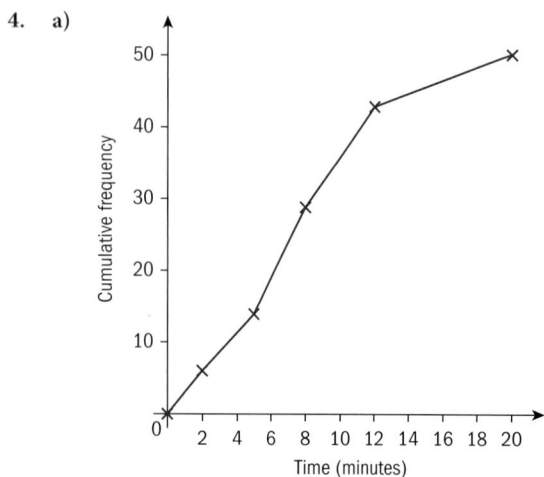

b) 7.3 minutes   **c)** 6 minutes

5. **i)** The trains were early.

ii)

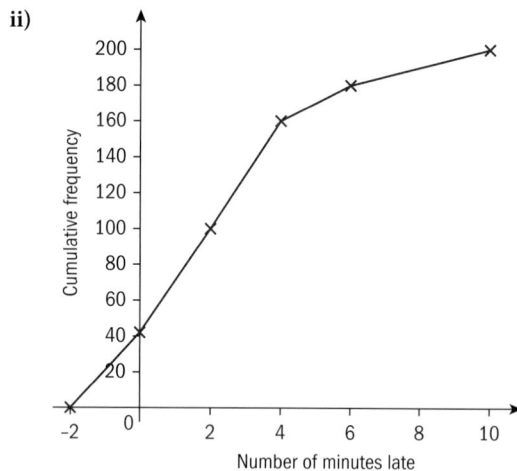

median = 2.2 minutes,
interquartile range = 3.4 minutes

6. **i)** $a = 494$, $b = 46$

ii)

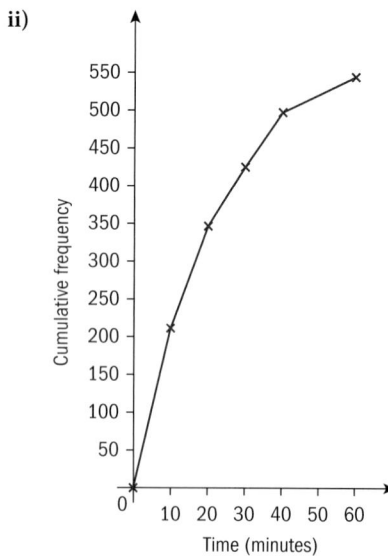

iii) 14.5

iv) $m = 18.2$ minutes (1 d.p.), $s = 14.2$ minutes
(1 d.p.)

v) 160

Exercise 3.5 page 53

1. **a)** A = 1.69 (2 d.p.), B = −1.62 (2 d.p.),
C = −1.04 (2 d.p.)

b) Group A

2. a) A = −0.07 (2 d.p.), B = 0.17 (2 d.p.), C = −0.21

b) Group C

c) Yes, but the signs are reversed for Groups B and C

3. a) 31.9375

b)

2	5 6 6 7 7 7 8 9 9
3	0 2 3 4
4	2 5
5	1

Key: 3|4 means 34

c) median = 29, lower quartile = 27, upper quartile = 33.5

d) 45 and 51 are outliers

e)

f) The distribution is positively skewed because there is a long tail to the right. Also, $Q_3 - Q_2 > Q_2 - Q_1$

4. a) median = 49.5 seconds, inter-quartile range = 58 seconds

b)

c) The distribution is positively skewed because there is a long tail to the right. However, the distribution is almost symmetrical between the quartiles.

d) A zero waiting time occurs when there is a train at the platform and the passenger does not have to wait.

Exercise 3.6 page 56

1. On average their pulse rates were higher after the PE lesson than after the mathematics lesson. They also varied more after the PE lesson than after the mathematics lesson.

2. a) mean = 1.1 kg, variance = 0.075 kg²

b) On average the new diet shows a higher weight loss than the traditional diet. The weight loss with the new diet is less varied than the weight loss with the traditional diet.

3. On average the plants are longer in garden A than in garden B. The overall range is slightly bigger for garden B than for garden A but the inter-quartile range is bigger for garden A than for garden B. This means that there is more variation in the central 50% of the data for garden A than for garden B.

Summary exercise 3 page 57

1. a)

b) The distribution of delays is positively skewed because there is a long tail to the right. Also, $Q_3 - Q_2 > Q_2 - Q_1$

c) On average there were longer delays on the second day than on the first. The delays on the second day were more varied than the delays on the first day.

2.

0	0 1 3 3 4 6 7 8 8
1	1 2 2 4 5
2	8
3	4

Key: 1|6 means 16 passengers

3. a) The data is continuous and has unequal class widths.

b) upper class boundary = 14.5, lower class boundary = 9.5

c)

d) mean =17.9 minutes (1 d.p.), standard deviation =12.2 minutes (1 d.p.)

e) the mean would decrease because the exit times for the 12 flights being excluded are all bigger than the current mean of 17.9 minutes

4. i) mean = 18.4 minutes (1 d.p.),
standard deviation = 13.3 minutes (1 d.p.)

ii)

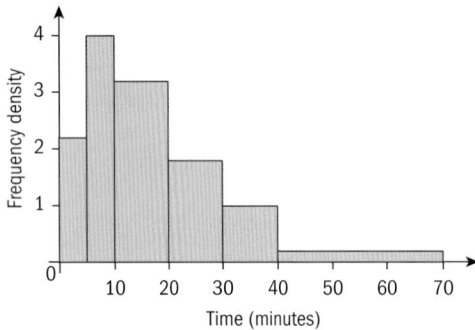

5. i) 40

ii)

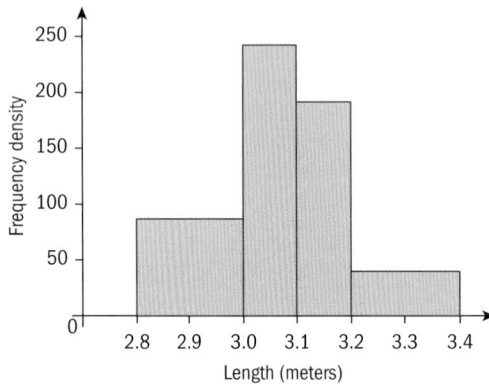

iii) $\frac{60}{68}$

6. i)

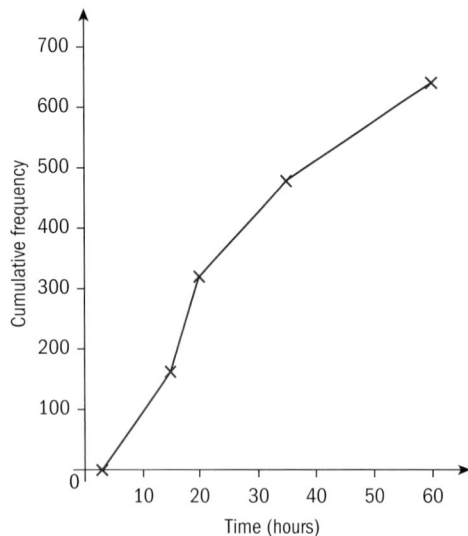

ii) Approximately 64

7. i) It puts the data in order / it shows the shape of the distribution but also keeps the detail of the data.

ii) median = 6.5, lower quartile = 5.4, upper quartile = 8.3

iii)

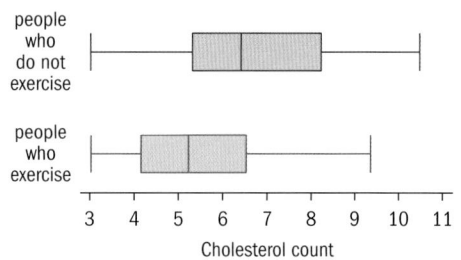

8. i) 30–35   **ii)** 24   **iii)** 110   **iv)** $\frac{24}{88}$

9. a)

0	4 5 6 7 7 7 7
1	2 3 8 8
2	6
3	3 3 4
4	4
5	6
6	9

Key: 2|4 means 24

b)

5	5 7 8
6	3
7	4
8	4 7 9
9	0 3 3 3 5 5 9

Key: 8|5 means 85%

c)

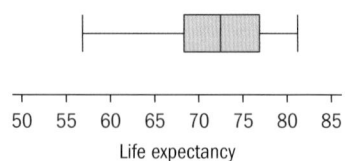

The overall distribution is negatively skewed but it is almost symmetrical between the quartiles.

d)

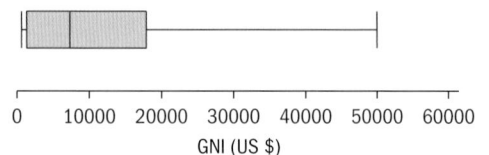

Positive skew.

e)

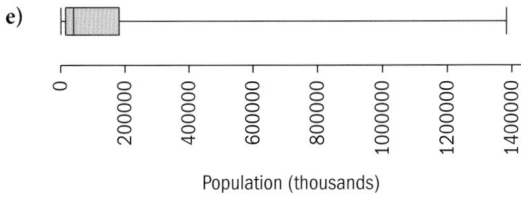

Population (thousands)

Very positively skewed.

10.

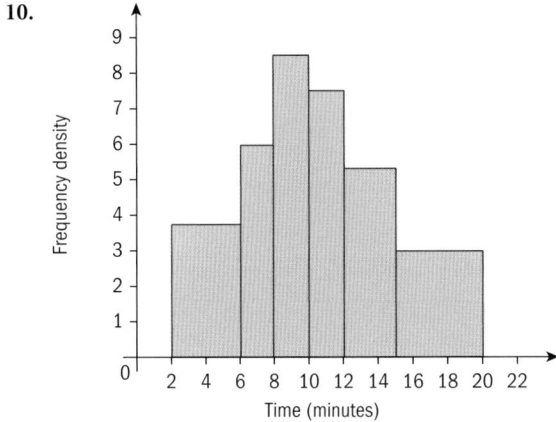

11.

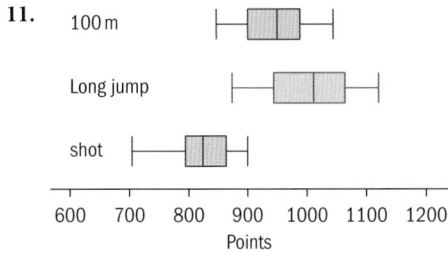

On average the points scored in the long jump are highest and the points scored in the shot are least. The variation in the points scored is greatest in the long jump and lowest in the shot.

Review exercise A page 63

1. **a)** 0   **b)** 25   **c)** 38.1

2. **a)** **i)** 1 and 4   **ii)** 3   **iii)** 3.4   **b)** 57

3. **a)** 67.1; 322.1

b)

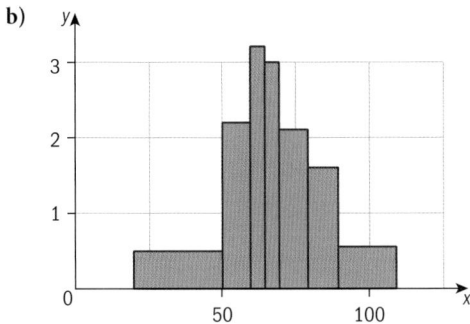

4. Mean = 6.25; Variance = 9.10

5. 35.6; 3.8   **6)** 5; 256.2   **7)** 146.8; 15.8

8. **a)** 19.6

   b) 60.08; 64.76; 63.14; 66.56; 68.54; 74.12; 71.78; 69.26

   c) 67.28 (2 d.p.)

9. **a)** 65.65; 103.13   **b)** 61 523.65 (2 d.p.)   **c)** 64.9

   d) The standard deviation of marks in the first class was just over 10, or nearly twice as big as the SD for the second class. On average the first class did slightly better (mean is about 2 marks higher) and their performance was much more variable than the second class.

10. **a)** 61.2; 15.79   **b)** $a = 4.44$ (2 d.p); $b = 1.14$ (2 d.p)

11.

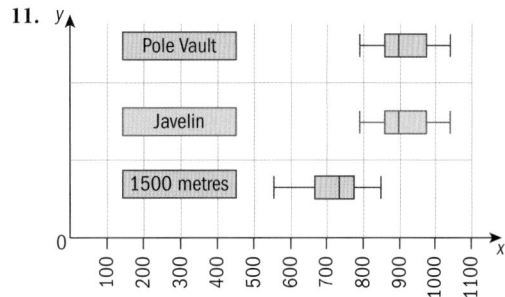

The points performances for the Pole Vault and Javelin are very similar – both in terms of average scores and the spread of performances. Performances in the 1500 metres (which is the last of the 10 events over two days) is very different – the median (average) score is much lower – some 150 points lower, and there is a long tail of poor performances (the graph is negatively skewed) compared with the other two disciplines.

12.

0	0 3 3 5 6 6 7
1	1 2 2 3 3 6 7 7 8
2	1 2 2 4 5
3	1

Key: 2|4 means 24 passengers

13. **a)**

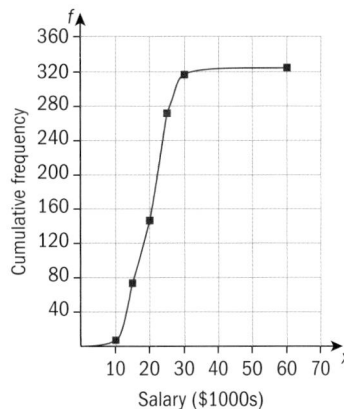

   b) $15 000; $21 000; $29 000   **c)** 3%

14. 0.850; 0.978 **15. i)** 41.39; 13.15 **ii)** 48; 13.44

16. i) 139; 83.1 **ii)** Team B; smaller standard deviation

17. i) 9.11 **ii)** 35412

18. i)

16 year olds		9 year olds
7, 4	11	
9, 8,	12	
7, 0	13	0, 2, 7,
8	14	2, 4,
	15	0, 1, 9,
5	16	0, 1, 4, 7,

Key: 7|13|2 means 13.7 minutes and 13.2 minutes

ii) 15.6

19. i) 54 min

ii)

0 1 2 3 4 5 6 time

20. i) $x = 15$; 0.75 m **ii)** 26.6 grams

21. i)

Median 270   pupils

ii) 160 **iii)** 500 **iv)** 268

22. i)

Flat screen		conventional
	6	5 7 9
6	7	1 4 5 7
9 5	8	5 6
6 4 2 1	9	
7 4	10	

Key 5|8|4 means 0.85 m for flat screen and 0.84 m for conventional

ii) 0.74; 0.13 **iii)** 0.927; 0.0882

23. i)

	14	3
	15	3 4 5
	16	1 4 8 8
	17	
	18	5

Key: |14|3 represents 14 300 dollars

ii) 15 400 **iii)** 0.121

24. i)

ii) 2.1 hours

25. i)

10	4 4 9
11	5 7
12	0 4 5
13	2 4
14	2 5
15	8
16	0 8

Key 10|4 represents 104

ii) 115; 125; 145

iii)

100 110 120 130 140 150 160 170

26. Two pie or bar charts

27. i)

0.5 20.5....ect....60.5 75.5   marks

ii) 37.5; 16.9

28. ii) a)

3	45
4	145
5	02
6	2
7	339
8	344556679
9	1

Key 3|4 rep 34, or stem width = 10

b) 79

4 Probability

Skills check page 70

1. a) $\frac{1}{6}$ b) $\frac{1}{6}$

2. HH, TT, HT, TH

Exercise 4.1 page 72

1. a) $A = \{1, 2, 4\}, B = \{1, 4\}, C = \{2, 3, 5\}, D = \{3, 6\}$

   b) $\frac{1}{2}, \frac{1}{3}, \frac{1}{2}, \frac{1}{3}$ c) $\{2\}$ d) $\frac{1}{6}$

   e) 0   f) $\frac{1}{2}$

2. a) A consonant is chosen.

   b) $A = \{A, E, I\}, B = \{B\}, C = \{A, B, C, D, E, G, I, M\},$
   $D = \{C, E\}$

   c) $\frac{1}{3}, \frac{1}{9}, \frac{8}{9}, \frac{2}{9}$ d) $\{A, E, I\}$

   e) $\frac{1}{3}$ f) $\frac{1}{9}$ g) $\frac{4}{9}$

3. c) 10

4. b) 5   d) 3

Exercise 4.2 page 74

1. {(V,V); (V,O); (V,C); (O,V); (O,O); (O,C); (C,V); (C,O); (C,C)}

2. {(V,O); (V,C); (O,V); (O,C); (C,V); (C,O)}

3.
Sum	1	2	3	4	5	6
1	X	3	4	5	6	7
2	3	X	5	6	7	8
3	4	5	X	7	8	9
4	5	6	7	X	9	10
5	6	7	8	9	X	11
6	7	8	9	10	11	X

   a) $\frac{2}{15}$ b) $\frac{1}{15}$ c) 0

4. a) {(H, 1); (H, 2); (H, 3); (H, 4); (H, 5); (H, 6); (T, 1); (T, 2); (T, 3); (T, 4); (T, 5); (T, 6)}

   b)
Score	1	2	3	4	5	6
H	2	3	4	5	6	7
T	3	4	5	6	7	8

5. a)
Product	1	2	3	4	5	6
1	1	2	3	4	5	6
2	2	4	6	8	10	12
3	3	6	9	12	15	18
4	4	8	12	16	20	24
5	5	10	15	20	25	30
6	6	12	18	24	30	36

   b) i) $\frac{1}{18}$ ii) $\frac{1}{12}$ iii) $\frac{1}{9}$ iv) $\frac{1}{12}$

6. a)
Low	1	2	3	4	5	6
1	1	1	1	1	1	1
2	1	2	2	2	2	2
3	1	2	3	3	3	3
4	1	2	3	4	4	4
5	1	2	3	4	5	5
6	1	2	3	4	5	6

   b) i) $\frac{7}{36}$ ii) $\frac{1}{12}$ iii) $\frac{1}{36}$

7. a)
Difference	1	2	3	4	5	6
1	0	1	2	3	4	5
2	1	0	1	2	3	4
3	2	1	0	1	2	3
4	3	2	1	0	1	2
5	4	3	2	1	0	1
6	5	4	3	2	1	0

   b) i) $\frac{1}{6}$ ii) $\frac{1}{18}$ iii) 0

Exercise 4.3 page 78

1. a) $\frac{15}{32}$ b) $\frac{15}{28}$

2. a) 0.4   b) 0.5

3.
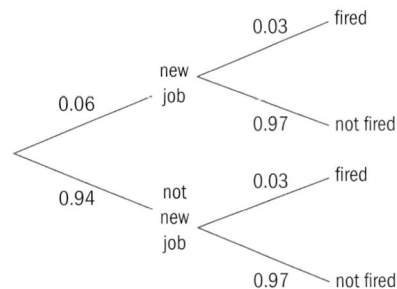

   a) 0.9118   b) 0.0864   c) 0.0018

4. a) $\frac{1}{8}$ b) $\frac{1}{4}$ c) $\frac{1}{4}$

5. a) $\frac{3}{10}$ b) $\frac{1}{30}$ c) $\frac{3}{10}$

Exercise 4.4 page 80

1. a)

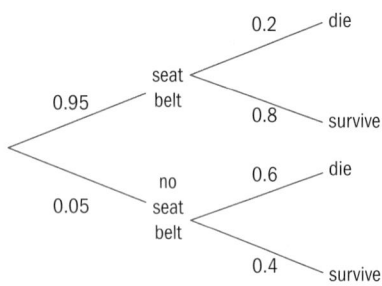

b) 0.02

2. a) $\dfrac{7}{300}$ **b)** $\dfrac{4}{7}$

3. a) 0.82 **b)** 0.492 **c)** 0.752

4. a) $\dfrac{130}{271}$ **b)** $\dfrac{49}{542}$ **c)** $\dfrac{49}{260}$ **d)** $\dfrac{153}{364}$

5. a) $\dfrac{25}{173}$ **b)** $\dfrac{7}{25}$

6.

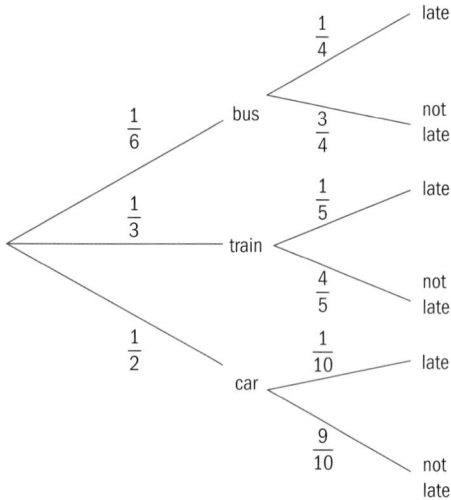

$\dfrac{19}{120}$

7. a) 0.035 **b)** 0.048 **c)** $\dfrac{35}{48}$

8. a)

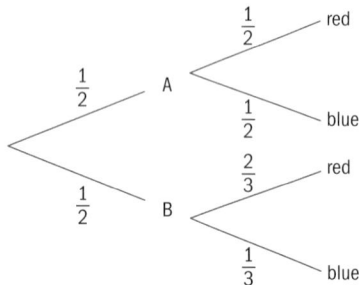

$\dfrac{7}{12}$

b) i) $\dfrac{25}{42}$ **ii)** $\dfrac{12}{25}$

Exercise 4.5 page 86

1. a) 0.28 **b)** 0.82 **c)** 0.12

4. a) 0.4 **b)** 0.5 **c)** 0.3

5. a) i) $\dfrac{59}{208}$ **ii)** $\dfrac{100}{208}$ **iii)** $\dfrac{59}{124}$

6. a) mutually exclusive **b)** mutually exclusive
c) neither **d)** mutually exclusive
e) independent **f)** mutually exclusive

7. a) true **b)** true
c) true **d)** false

Summary exercise 4 page 87

1. a)

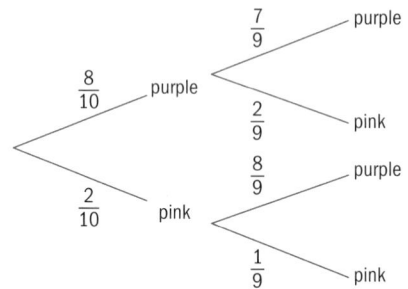

b) $\dfrac{4}{5}$ **c)** $\dfrac{7}{9}$

2. a) 0.002978 **b)** 0.671 (3 s.f.)
c) There is more than a 67% chance that a person who has tested positive does not have the disease.

3. a) 0 **b)** $\dfrac{5}{6}$
c) A and B are not independent since $P(A \mid B) \neq P(A)$

4. a) i) $\dfrac{1}{4}$ **ii)** $\dfrac{1}{4}$ **iii)** $\dfrac{2}{5}$
b) i) no, since $P(A \cap B) \neq 0$
ii) no, since $P(A \mid B) \neq P(A)$

5. a)

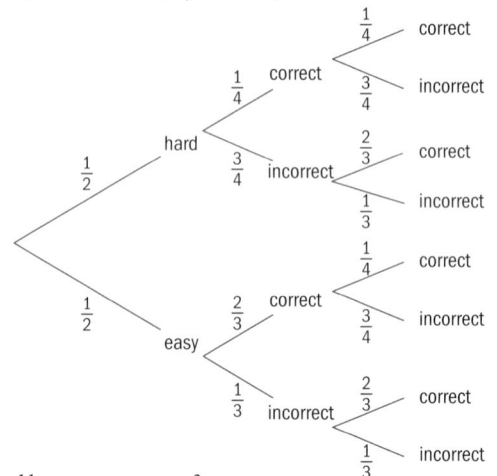

b) $\dfrac{11}{96}$ **c)** $\dfrac{3}{11}$

6. **a)**

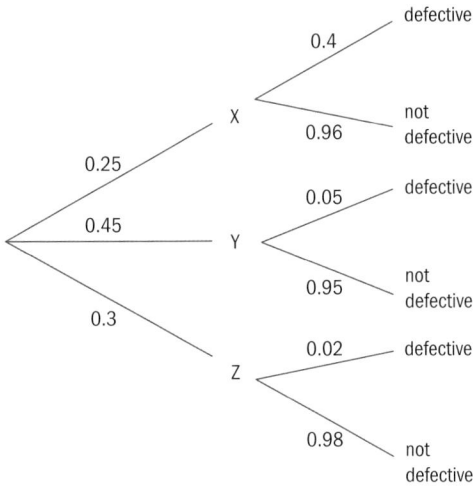

b) **i)** 0.4275 **ii)** 0.9615

c) $\dfrac{285}{641}$ or 0.445 (3 s.f.)

7. **a)** 0.35

c) The golfer's form from the first week is likely to be carried into the second week.

8. **a)** $\dfrac{1}{6}$ **b)** $\dfrac{1}{3}$ **c)** $\dfrac{1}{2}$

9. 0

10. **a)** 0.05 **b)** 0.125 **c)** 0.8

11. **a)** $P(A \cap B) = P(A) \times P(B)$, $P(A \cup B)$
$= P(A) + P(B) - P(A) \times P(B)$

b) $\dfrac{2}{3}$ or $\dfrac{1}{2}$

12. **a)** 0.2 **b)** $\dfrac{2}{7}$

13. **i)** 0.8 **ii)** 0.625

14. **i)** $\dfrac{618}{1281}$ **ii)** $\dfrac{412}{1281}$

iii)

$P(E/M) = \dfrac{412}{618} = \dfrac{2}{3}$, $P(E) = \dfrac{717}{1281} = \dfrac{239}{427}$, $P(E \mid M) \neq P(E)$

Therefore M and E are not independent events.

iv) $\dfrac{358}{564}$

15. **ii)** 0.7807 **iii)** 0.372 (3 s.f.)

5 Probability distributions and discrete random variables

Skills check page 92

1. 12, 6, 4, 3

2. $a = 0.1, b = 0.3$

Exercise 5.1 page 94

1. **a)** not a discrete random variable since $\sum P(X = x) \neq 1$

b) a discrete random variable

c) a discrete random variable

d) not a discrete random variable since $P(X = 5) < 0$

2. **a)**

x	1	2	3	4	5	6
$P(X = x)$	$\dfrac{1}{6}$	$\dfrac{1}{6}$	$\dfrac{1}{6}$	$\dfrac{1}{6}$	$\dfrac{1}{6}$	$\dfrac{1}{6}$

b)

y	2	4	6	8	10	12
$P(Y = y)$	$\dfrac{1}{6}$	$\dfrac{1}{6}$	$\dfrac{1}{6}$	$\dfrac{1}{6}$	$\dfrac{1}{6}$	$\dfrac{1}{6}$

c)

z	1	4	9	16	25	36
$P(Z = z)$	$\dfrac{1}{6}$	$\dfrac{1}{6}$	$\dfrac{1}{6}$	$\dfrac{1}{6}$	$\dfrac{1}{6}$	$\dfrac{1}{6}$

d)

w	0	1
$P(W = w)$	$\dfrac{2}{3}$	$\dfrac{1}{3}$

3.

x	0	1	2
$P(X = x)$	$\dfrac{1}{4}$	$\dfrac{1}{2}$	$\dfrac{1}{4}$

4. **a)** **i)** 0.3 **ii)** 0.8

b) **i)** $\dfrac{1}{6}$ **ii)** $\dfrac{5}{6}$

c) **i)** 0 **ii)** 0.1

Exercise 5.2 page 96

1. **a)**

r	1	2	3	4	5
Probability	$\dfrac{1}{15}$	$\dfrac{2}{15}$	$\dfrac{3}{15}$	$\dfrac{4}{15}$	$\dfrac{5}{15}$

b)

r	1	2	3	4
Probability	$\dfrac{12}{25}$	$\dfrac{6}{25}$	$\dfrac{4}{25}$	$\dfrac{3}{25}$

c) not a probability function since it is undefined for $r = 1$

d)

r	1	2	3	4
Probability	$\dfrac{30}{77}$	$\dfrac{20}{77}$	$\dfrac{15}{77}$	$\dfrac{12}{77}$

2. a)

z	1	2	3	4
Probability	$\frac{4}{10}$	$\frac{3}{10}$	$\frac{2}{10}$	$\frac{1}{10}$

b)

y	1	2	3	4	5
Probability	$\frac{1}{5}$	$\frac{1}{5}$	$\frac{1}{5}$	$\frac{1}{5}$	$\frac{1}{5}$

c)

w	2	3	4	5	6	7	8	9	10	11	12
Probability	$\frac{1}{36}$	$\frac{2}{36}$	$\frac{3}{36}$	$\frac{4}{36}$	$\frac{5}{36}$	$\frac{6}{36}$	$\frac{5}{36}$	$\frac{4}{36}$	$\frac{3}{36}$	$\frac{2}{36}$	$\frac{1}{36}$

d)

r	1	2	3	4
Probability	$\frac{1}{16}$	$\frac{3}{16}$	$\frac{5}{16}$	$\frac{7}{16}$

3. a) $\frac{12}{15}$ **b)** $\frac{6}{15}$ **c)** $\frac{3}{15}$

d) $\frac{1}{2}$ **e)** $\frac{1}{4}$

4. b) i) 0.52   **ii)** $\frac{3}{13}$

5. $c = \frac{1}{30}$, $\Pr\{Y < 3\} = \frac{5}{30}$

Exercise 5.3 page 99

1. a) 3.2   **b)** −0.2   **c)** 7.6875

2. a) 2   **b)** 3

3. a) i) 0.3   **ii)** 2.7   **iii)** 9.3

b) i) $\frac{1}{6}$ **ii)** 7.5   **iii)** 61.5

4. $\frac{161}{36}$ **5.** $a = 0.4, b = 0.2$

6. $a = 0.175, b = 0.125, P(X > E(X)) = 0.325$

Exercise 5.4 page 102

1. a) $E(X) = 7.1, \text{Var}(X) = 1.29$

b) $E(X) = 0.1, \text{Var}(X) = 1.29$

c) $E(X) = 1.9375, \text{Var}(X) = 1.43$ (3 s.f.)

2. a) mean = 2.33 (3 s.f.), variance = 1.56 (3 s.f.)

b) mean = 3.5, variance = 2.92 (3 s.f.)

c) mean = 7, variance = 5.83 (3 s.f.)

3. a) i) 0.3 **ii)** $E(X) = 2.9, \text{Var}(X) = 1.69$

b) i) $\frac{1}{6}$ **ii)** $E(X) = 4.5, \text{Var}(X) = 0.917$ (3 s.f.)

4. $E(Y) = 2.53$ (3 s.f.), $\text{Var}(Y) = 1.97$ (3 s.f.)

5. $a = 0.1, b = 0.5, \text{Var}(X) = 2.21$

6. $a = 0.2, b = 0.3, \text{Var}(X) = 18.01$

Summary exercise 5 page 103

1. b) $\frac{18}{25}$ **c)** 1.92

2. b) 5.5   **c)** 2.42 (3 s.f.)

3. a) 0.4   **b)** 6.56   **c)** 0

4. X is not a discrete uniform distribution, Y and Z are discrete uniform distributions.

5. a) 0.3   **b)** 8   **c)** 1

d) −11   **e)** 4

6. a) i) mean = 2, standard deviation = 1.14 (3 s.f.)

ii) 5.1 minutes

b) i) 0.35   **ii)** $\frac{1}{14}$

7. a) $a + b = 0.4, 8a + 11b = 3.95$ **b)** $a = 0.15, b = 0.25$

c) 2.0475

8. a)

x	1	2	3	4	5
P(X = x)	$\frac{9}{25}$	$\frac{7}{25}$	$\frac{5}{25}$	$\frac{3}{25}$	$\frac{1}{25}$

b) $\frac{8}{25}$ **c)** 2.2   **d)** 1.36

9. a)

x	10	12	15	16	18	20	24	25
P(X = x)	0.1	0.2	0.05	0.05	0.2	0.1	0.2	0.1

b) 17.85   **c)** 26.7275

10. b) $E(X) = 3.33$ (3 s.f.)   **c)** 0.689 (3 s.f.)

11. a) $\frac{1}{28}$ **b)** $E(X) = 5, \text{Var}(X) = 3$

c) $\frac{3}{28}$ **d)** mean = 3500, variance = 6 750 000

12. i) $\frac{1}{3}$ **iii)** $\frac{326}{81}$ or 4.02 (3 s.f.)

iv) $\frac{2}{27}$ **v)** $\frac{1}{6}$

13. i) $\frac{1280}{6561}$ or 0.195 (3 s.f.)

ii)

x	2	4	6	7	8	9	10	11	12
P(X = x)	$\frac{1}{36}$	$\frac{2}{36}$	$\frac{5}{36}$	$\frac{4}{36}$	$\frac{4}{36}$	$\frac{4}{36}$	$\frac{4}{36}$	$\frac{8}{36}$	$\frac{4}{36}$

iii) $\frac{26}{3}$ or 8.67 (3 s.f.)   **iv)** $\frac{20}{36}$

14. i) $2

iii)

Profit ($)	4	2	0	−1
Probability	0.2	0.288	0.184	0.328

iv) $1.05 (3 s.f.)

15. $a = 0.2, b = 0.25$

16. i) The cost of the roller coaster and the cost of the water slide must equal the mean since the standard deviation is zero.

   ii) $1.03 (3 s.f.)

17. i) $P(X = 1) = \frac{1}{14}$ **ii)** $\frac{15}{28}$

     $P(X = 2) = \frac{3}{7}$

     $P(X = 3) = \frac{3}{7}$

     $P(X = 4) = \frac{1}{14}$

18.

x	0	1	2	3
$P(X = x)$	$\frac{7}{24}$	$\frac{21}{40}$	$\frac{7}{40}$	$\frac{1}{120}$

6 Permutations and combinations

Skills check page 108

1. HH, TT, HT, TH

2. H1, H2, H3, H4, H5, H6, T1, T2, T3, T4, T5, T6

Exercise 6.1 page 110

1. a) 720   **b)** 6   **c)** 479 001 600

2. a) 7   **b)** 20   **c)** 1

3. 24   **4.** 40 320   **5.** 40 320

6. 120   **7.** 120   **8.** 3 628 800

Exercise 6.2 page 111

1. a) 8   **b)** 210   **c)** 120   **d)** 120

2. a) 13 366 080   **b)** 2.17×10^{11}

3. 15120   **4.** 3024   **5.** 1 814 400

6. 6720   **7.** 336   **8.** 360

Exercise 6.3 page 113

1. 240

2. a) 40 320   **b)** 2880

3. a) 241 920   **b)** 40 320

4. 96   **5.** 24   **6.** 42

7. a) 72   **b)** 24

8. a) 261 273 600   **b)** 3 110 400

9. 144

Exercise 6.4 page 115

1. a) 10 080   **b)** 360

2. a) 120   **b)** 720

3. 1680   **4.** 56

5. 1960   **6.** 8008

7. a) 151 200   **b)** 5040

Exercise 6.5 page 118

1. 27 405   **2.** 15 504

3. b) Choosing r objects to assign to the 'taken' group has a one-to-one correspondence with choosing the $n - r$ objects to assign to the 'left behind' group.

4. a) 14 950   **b)** 8965

5. 53 130

6. 1771; they need to ensure that neither the winner nor the runner up test positive for performance-enhancing drugs

Exercise 6.6 page 119

1. $\frac{2}{5}$ **2.** $\frac{5}{42}$

3. a) $\frac{2}{11}$ **b)** $\frac{1}{77}$ **c)** $\frac{27}{154}$

4. 0.398   **5.** 0.0128   **6.** $\frac{1}{105}$

Summary exercise 6 page 120

1. 369 600   **2.** $\frac{1}{945}$

3. 0.4   **4.** $\frac{1}{3}$

5. i) 2 177 280   **ii)** 90

6. a) 1260 **b) i)** 1680   **ii)** 360

   c) 4920

7. a) i) 645 120 **ii)** 50 803 200

   b) i) 21   **ii)** 6   **iii)** 51

8. i) 120   **ii)** 186   **iii)** 90

9. i) 90 720   **ii)** 720

10. i) 259 459 200 **ii)** 3 628 800 **iii)** $\frac{141}{143}$

Review exercise B page 123

1 $\frac{29}{36}$

2 a) 0.5   **b)** 0.2   **c)** 0.8   **d)** 0.8

3 a) 0.3   **b)** $\frac{2}{3}$

4 i) 0.503 **ii)** 0.259 **iii)** Yes **iv)** 0.54

5 a) $a = 0.3$ **b)** 24.96   **c)** 0.2

6 a) $4a + 7b = 2.5; a + b = 0.4$

   b) $a = 0.1; b = 0.3$ **c)** 1.65

7 a)

1	2	3	4
$\frac{1}{8}$	$\frac{5}{24}$	$\frac{7}{24}$	$\frac{3}{8}$

   b) 0.5   **c)** $\frac{35}{12}$ **d)** $\frac{155}{144}$

8 a) Proof **b)** $\frac{49}{18}$; Proof **c)** $\frac{83}{324}$

9 90 000

10. a) 3 628 806   **b)** 3 628 800

11. a) 7 893 600   **b)** 7 893 489

12. i) $\dfrac{3}{8}$ **ii)** 0.405   **iii)** $\dfrac{10}{17}$

13. i) $\dfrac{12}{36}$ **ii)** $\dfrac{20}{36}$

    iii) Not mutually exclusive

14. i) 0.252   **ii)** 0.440

15. i) 33 033 000 (33 000 000)   **ii)** 86 400

    iii) 288

16. i) $\dfrac{1}{2}$

    ii)

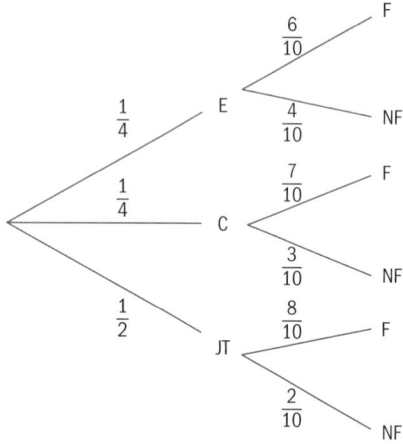

    iii) 0.725   **iv)** 0.273

17. i)

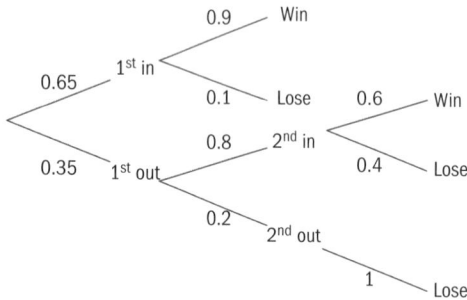

    ii) 0.247   **iii)** 0.263

18. i) 0.117   **ii)** 0.783   **iii)** 0.851

    iv)

X	0	1	2
P(X = x)	$\dfrac{3}{60}$	$\dfrac{17}{60}$	$\dfrac{40}{60}$

19. i) 0.273   **ii)** $\dfrac{12}{55}$

    iii)

x	0	1	2	3
P(X = x)	$\dfrac{14}{55}$	$\dfrac{28}{55}$	$\dfrac{12}{55}$	$\dfrac{1}{55}$
decimal	0.255	0.509	0.218	0.018

20. i) $q = 0.15$ **ii)** 1.41

21. i) P (odd) = 0.4   **ii)** 0.3

    iii)

l	3	4	5
P(L = l)	0.1	0.3	0.6

    iv) 4.5; 0.45

22. i)

X	1	2	3	4	5	6
P(X = x)	$\dfrac{11}{36}$	$\dfrac{9}{36}$	$\dfrac{7}{36}$	$\dfrac{5}{36}$	$\dfrac{3}{36}$	$\dfrac{1}{36}$

    ii) $\dfrac{91}{36}$

23. i)

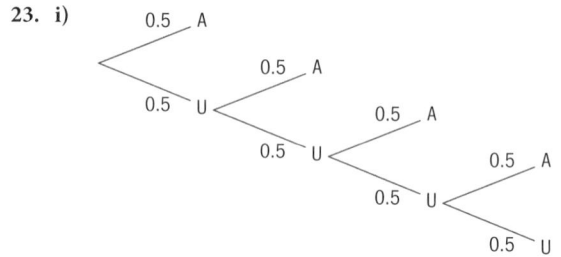

    ii)

X	0	1	2	3	4
P(X = x)	$\dfrac{1}{2}$	$\dfrac{1}{4}$	$\dfrac{1}{8}$	$\dfrac{1}{16}$	$\dfrac{1}{16}$

    iii) $\dfrac{15}{16}$

24. 37.5

25. i) 0.486   **ii)** 5.33   **iii)** $a = 1$

26. i) Proof

    ii)

X	0	1	2
P(X = x)	$\dfrac{7}{15}$	$\dfrac{7}{15}$	$\dfrac{7}{15}$

    iii) $\dfrac{3}{5}$

27. i) Proof

    ii)

X	120	60	40	30
P(X = x)	$\dfrac{1}{45}$	$\dfrac{2}{45}$	$\dfrac{3}{45}$	$\dfrac{4}{45}$

24	20	17.14	15	13.3
$\dfrac{5}{45}$	$\dfrac{6}{45}$	$\dfrac{7}{45}$	$\dfrac{8}{45}$	$\dfrac{9}{45}$

    iii) 13.3   **iv)** 0.444

28. i) 362 880   **ii)** 151 200   **iii)** 64

29. a) i) 24   **ii)** 28   **b)** $\dfrac{2}{3}$

30. i) 360   **ii)** 12   **iii)** 360

    iv) 720   **v)** 1170

31. a) i) 19 958 400   **ii)** 362 880

     b) i) 93 024   **ii)** 31 104

32. i) 831 600   **ii)** 900   **iii)** 126

33. i) $a + b = 0.45$ **ii)** $a = 0.15; b = 0.3$

34. i) Proof   **ii)** $\dfrac{5}{72}$

35. i) Proof   **ii)** 162   **iii)** 688 747 536

36. a) i) 18 564   **ii)** 6188

     b) i) 40 320   **ii)** 2880

37. i) 0.0556   **ii)** 2.78; 1.17   **iii)** 0.611

38. i) 829 440   **ii)** 2 438 553 600 **iii)** 560

39. 0.607

40. i) 13 580   **ii)** 288   **iii)** 240

41. i) Proof   **ii)** Proof

     iii)

X	1	2	3	4	5	6	7	8
Prob		$\dfrac{5}{32}$		$\dfrac{7}{32}$		$\dfrac{3}{32}$		

     iv) Exclusive

42. i) $x = 0.438$ **ii)** 0.3

43. i) 0.072   **ii)** 0.25

7 The binomial distribution

Skills check page 134

1. a) 3 628 800   **b)** 6435

2. 10.4

Exercise 7.1 page 139

1. a) 1   7   21   35   35   21   7   1

     b) i) 35   **ii)** 21

2. a) i) 210   **ii)** 1   **iii)** 5005

     iv) 4950

     b) i) 45   **ii)** 462   **iii)** 1

     iv) 4.71×10^{13} (3 s.f.)

3. a) 0.234 (3 s.f.)   **b)** 0.0938 (3 s.f.)

4. a) 0.324 (3 s.f.)   **b)** 0.0102 (3 s.f.)

5. a) 0.142 (3 s.f.)   **b)** 0.0420 (3 s.f.)

6. a) 0.00149 (3 s.f.)   **b)** 0.231 (3 s.f.)

7. a) 0.0173 (3 s.f.)   **b)** 0.137 (3 s.f.)

8. a) 0.0131 (3 s.f.)   **b)** 0.0393 (3 s.f.)

9. a) 0.234 (3 s.f.)   **b)** 0.0938 (3 s.f.)

10. a) i) 0.000977 (3 s.f.)   **ii)** 0.00977 (3 s.f.)

       iii) 0.0439 (3 s.f.)   **iv)** 0.117 (3 s.f.)

     b) i) 0.0107 (3 s.f.)   **ii)** 0.945 (3 s.f.)

       iii) 0.828 (3 s.f.)

11. a) 0.264 (3 s.f.)   **b)** 1.00 (3 s.f.)

12. a) 0.0913 (3 s.f.)   **b)** 9.09×10^{-13} (3 s.f.)

13. a) 0.237 (3 s.f.)   **b)** 0.0879 (3 s.f.)

Exercise 7.2 page 141

1. a) 17.5   **b)** 8.75

2. mean = 30, standard deviation = 4.24 (3 s.f.)

3. a) 2   **b)** 0.421 (3 s.f.)

4. 100

5. a) mean = 16, variance = 3.2   **b)** 0.589 (3 s.f.)

6. $\text{Var}(X) = 20p(1 - p), p = 0.5$

7. 0.360 (3 s.f.)

Exercise 7.3 page 146

1. a) No, there are not a fixed number of trials.

     b) Yes, $n = 10, p = \dfrac{1}{6}$

     c) No, since the balls are taken without replacement the trials are not independent.

     d) Yes, $n = 5, p = \dfrac{1}{2}$

     e) Yes, $n = 25, p = \dfrac{1}{6}$

     f) No, you are not counting the number of times a particular outcome is observed.

2. a) i) 0.0424 (3 s.f.)   **ii)** 0.141 (3 s.f.)

       iii) 0.228 (3 s.f.)   **iv)** 0.236 (3 s.f.)

     b) 3   **c)** mean = 3, variance = 2.7

3. a) The cars are independent. $n = 40, p = 0.08$

     b) The matches are independent. $n = 48, p = 0.02$

     c) $n = 50, p = 0.3$

     d) There are a sufficient number of balls to assume that the colours of the balls are independent. $n = 50, p = 0.3$

4. a) mean = 3, variance = 1.5

     b) mean = 3, variance = 3.07 (3 s.f.)

     c) No, the variance of 3.07 is more than double the variance given by the binomial model.

Summary exercise 7 page 147

1. a) 0.817 (3 s.f.)   **b)** 0.408 (3 s.f.)   **c)** 13

2. a) 125   **b)** 1.70 (3 s.f.)

3. a) 0.825 (3 s.f.)   **b)** 0.177 (3 s.f.)

4. a) i) 0.683 (3 s.f.)   **ii)** 0.576

     b) 0.166 (3 s.f.)

     c) mean = 2, standard deviation = 1.10 (3 s.f.)

d) i) mean = 2.1, standard deviation = 1.60 (3 s.f.)

ii) These values do not support Louise's belief. The mean of 2.1 is close to the mean given by the binomial model but the standard deviation of 1.60 is approximately 45% greater than that given by the binomial model.

5. a) i) 0.839 (3 s.f.) **ii)** 0.187 (3 s.f.)

iii) 0.835 (3 s.f.)

b) mean = 210, variance = 151.2

6. a) i) 0.219 (3 s.f.) **ii)** 0.950 (3 s.f.)

iii) 0 **iv)** 0.686 (3 s.f.)

b) i) mean = 9, standard deviation = 1.90 (3 s.f.)

ii) mean = 9.1, standard deviation = 1.58 (3 s.f.)

These values are close to those given by the binomial model and support Ronnie's claim.

7. a) i) 0.839 (3 s.f.) **ii)** 0.107 (3 s.f.)

iii) 0.831 (3 s.f.)

b) For the binomial model the mean is 8 and the variance is 6.4. These values do not support Conn's claim because of the high value of his variance.

8. a) 0.873 (3 s.f.) **b)** 0.0266 (3 s.f.)

9. a) Two assumptions from: there are a fixed number of trials; each trial must have the same two possible outcomes; the outcomes of the trials are independent of each other; the probability of the outcomes remains constant.

b) i) 0.0402 (3 s.f.) **ii)** 0.314 (3 s.f.)

c) Proof

10. a) 0.411 (3 s.f.) **b)** 0.210 (3 s.f.)

c) The colours of the beads are independent.

11. i) 0.994 (3 s.f.) **ii)** 0.405 (3 s.f.)

12. i) 0.264 (3 s.f.) **ii)** $2830 (3 s.f.)

13. 0.655 (3 s.f.)

8 The normal distribution

Skills check page 152

1. $\mu = 41$, $\sigma = 10$

2. 5

Exercise 8.1 page 156

1. a) i) 2 **ii)** −0.5 **iii)** 1.5 **iv)** 0

b) i) 65.1 **ii)** 39.2 **iii)** 53.2 **iv)** 70

2. a) i) −1.4 **ii)** −5.6 **iii)** 0.86 **iv)** −0.06

b) i) 98.5 **ii)** 76.5 **iii)** 84 **iv)** 92

3. a) i) 1 **ii)** −0.5 **iii)** −2.5

iv) 0.233 (3 dp)

b) i) 11.4 **ii)** −12.6 **iii)** 0.6 **iv)** 24.6

4. a) 6 **b)** 59

5. 7.5

6. 529.5

Exercise 8.2 page 161

1. a) 0.4761 **b)** 0.9957 **c)** 0.7881

d) 0.1089 **e)** 0.7721

2. a) 0.8599 **b)** 0.6179 **c)** 0.2358

d) 0.0905

3. a) 0.2061 **b)** 0.7675 **c)** 0.8371

4. a) 0.9281 **b)** 0.4716 **c)** 0.0205

d) 0.9638

5. a) 1.282 **b)** 2.576 **c)** −1.960

d) −0.842 **e)** 1.555 **f)** −2.120

Exercise 8.3 page 165

1. a) 0.9641 **b)** 0.2119 **c)** 0.1587

2. a) 0.3085 **b)** 0.9599 **c)** 0.3354

3. a) 0.9772 **b)** 0.3694 **c)** 0.3596

4. a) 0.5 **b)** 0.4279 **c)** 0.6826

5. a) 0.9873 **b)** 0.2881 **c)** 0.6170

6. a) 31.6 **b)** 26.0

7. a) 93.7 **b)** 80.8 **c)** 86.7

8. a) 4.96 **b)** 2.03

9. 15.2

10. 2.71

11. 5.22

12. 4.98

13. 0.546

14. $\mu = 31.9$, $\sigma = 5.86$

15. $\mu = 73.3$, $\sigma = 18.7$

16. $\mu = 1.54$, $\sigma = 2.21$

17. $\mu = 1074$, $\sigma = 69.4$

18. $\mu = 47.3$, $\sigma = 5.58$

Exercise 8.4 page 170

1. 124.7

2. a) 0.2798 **b)** 0.4979

3. a) Proof **b)** 45.92% **c)** 0.1582

4. $\mu = 29.4$, $\sigma = 8.29$

5. **a)** 0.7734 **b)** 48.1 hours **c)** 49.6 hours

6. **a)** 0.1056 **b)** 0.8640 **c)** 343 ml
d) 333 ml **e)** 6.08 ml

7. **a)** 0.2266 **b)** 0.1056

8. **a)** **i)** 11.51% **ii)** 86.21% **iii)** 483 seconds
b) 351 seconds

9. **a)** 662 ml **b)** 0.0002%

10. **a)** 2.28% **b)** 29 **c)** 17.95 mm
d) 0.0063

Summary exercise 8 page 172

1. **a)** 0.0228 **b)** 0.2384 **c)** 87.4

2. **a)** 0.0196 **b)** 0.0530 **c)** 0.0010
d) It is unlikely that the performances in the two events are independent.

3. **a)** 0.1587 **b)** 0.1624 **c)** 3.88 ml

4. **b)** $\mu = 291$ hours, $\sigma = 29.7$ hours **c)** £645.38

5. **a)** 0.1186 **b)** 0.7437 **c)** 265

6. 9.45 minutes

7. **a)** 0.0228 **b)** 82 **c)** 0.0150
d) It is unlikely that the performance in the written paper and the performance in the project are independent of one another.

8. **a)** **i)** Proof **ii)** 85 **b)** 0.2684

9. **a)** **i)** 0.2266 **ii)** 0.6678 **iii)** 0.0287
b) profit = £34.75

10. **i)** Weights, heights etc. **ii)** $\mu = 12.9$ **iii)** 7

11. **i)** 0.256 **ii)** 1.71 and 2.09

12. **a)** $\sigma = 0.545$; $\mu = 0.693$ **b)** $a = 3.09$

13. **i)** $\mu = 9.9$; $\sigma = 3.15$ or 3.16 **ii)** 317

14. 0.212

15. **i)** $\sigma = 7.29$; **ii)** 0.136 **iii)** 0.370

16. **a)** 0.0228 **b)** $\sigma = 0.0323$

17. **a)** **i)** 0.226 **ii)** 0.668 **iii)** 0.0287
b) 0.1009

9 The normal approximation to the binomial distribution

Skills check page 178

1. mean = 2.4, variance = 1.68

2. **a)** 0.9772 **b)** 0.3694 **c)** 0.3596

Exercise 9.1 page 179

1.

X	0	1	2	3	4	5	6
probability	0.015625	0.09375	0.234375	0.3125	0.234375	0.09375	0.015625

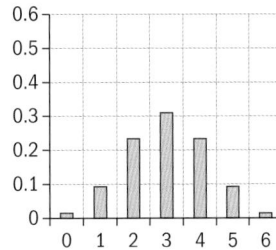

2.

X	0	1	2	3	4	5	6
probability	0.531441	0.354294	0.098415	0.01458	0.001215	5.4E-05	0.000001

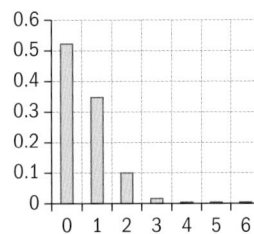

Exercise 9.2 page 181

1. **a)** $P(Y < 41.5)$ **b)** $P(Y > 31.5)$ **c)** $P(Y > 8.5)$
d) $P(42.5 < Y < 85.5)$

2.

X	0	1	2	3	4	5	6
probability	0.015625	0.09375	0.234375	0.3125	0.234375	0.09375	0.015625

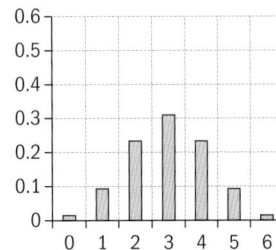

X	0	1	2	3	4	5	6
probability	0.531441	0.354294	0.098415	0.01458	0.001215	5.4E-05	0.000001

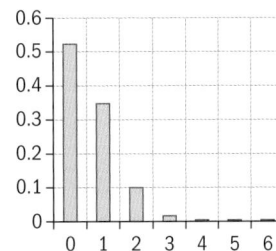

Exercise 9.3 page 183

1. **a)** N(35, 10.5) **c)** N(100, 80)
2. **a)** 0.9776 **b)** 0.8864 **c)** 0.5113
   d) 0.8263 **e)** 0.6289
3. **a)** **i)** 0.6923 **ii)** 0.6973
   b) **i)** 0.005 **ii)** 0.72%
4. $np > 5$ and $n(1 - p) > 5$, N(np, $np(1 - p)$)
5. **a)** 0.2342 **b)** 0.0268
6. **a)** B(50, 0.25) **b)** 0.0139 **c)** 0.8023
7. **a)** 0.8688
   b) they are unlikely to be a random sample since they are patients who are more likely to be ill and to have raised blood pressure

Summary exercise 9 page 184

1. **a)** 3.2 **b)** 0.589
2. **a)** 0.0547 **b)** 25 **c)** 250
3. **a)** 0.1252 **b)** 16.66 **c)** 166.66
4. **a)** **i)** 0.850 **ii)** 0.350
   b) 62.275 **c)** 0.425
5. **a)** 0.805 **b)** 44.77
6. **a)** **i)** 0.0718 **ii)** 0.729 **iii)** 0.967
   b) 22.5
7. **i)** 0.748 **ii)** 0.887
8. **i)** 0.00563 **ii)** 0.526 **iii)** 0.956
9. 0.794 **10. i)** 0.0779 **ii)** 0.275
11. 0.677
12. **i)** 5080 **ii)** 0.0273 **iii)** 0.730
13. **i)** 0.398 **ii)** $n = 9$ **iii)** 0.972
14. 0.704

Review exercise C page 187

1. **i)** 0.209 **ii)** 0.124
2. **i)** 0.109 **ii)** 1.00
3. **a)** 1.597 **b)** 0.405
4. **a)** 0.774 **b)** 0.405 **c)** 27
5. **a)** 0.145 **b)** Group AB, p = 0.12, q = 0.88, n = 500
   c) 0.480
6. **a)** 0.977 **b)** 0.775 **c)** a = 164
7. **a)** 0.048 **b)** 0.050 **c)** 1.94 ml
8. **a)** 2.03 mm **b)** 0.70 % **c)** 0.0075
9. 29.4 mm **10.** 0.71
11. **a)** Proof **b)** 166 **c)** 0.087
12. **i)** 0.0584 **ii)** 0.307 **iii)** 0.829
13. **i)** 0.334 **ii)** 49.9

14. $\sigma = 6.70$ $\mu = 23.0$
15. **i)** 0.0276 **ii)** 7.72
16. **a)** $s = 7.24$ or 7.23 **b)** 546
17. **i)** 0.311 **ii)** Not possible **iii)** 0.181
18. $\sigma = 8.91$; $\mu = 23.6$
19. **i)** 0.774 **ii)** 0.204 **iii)** 0.0451

Paper A page 192

1. $\bar{x} = 30.024$; $\sigma = 0.461$
2.

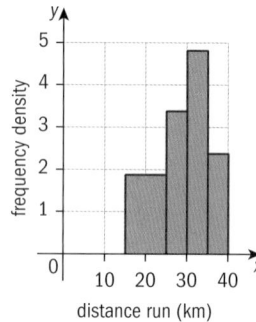

distance run (km)

3. **i)** $\sigma = 0.950$ **ii)** 0.4164
4. **i)** 0.159 **ii)** 0.579
5. **i)** $\dfrac{12}{134}$ **ii)** $\dfrac{15}{67}$ **iii)** $\dfrac{24}{47}$
6. **a)** **i)** 117 **ii)** 71 **iii)** 270 **b)** 30
7. **i)** 0.315
   ii)

X	1	2	3	4	5	6	8	9	10	12	15	16	20	25
P(/36)	1	2	2	3	2	2	2	1	2	2	2	1	2	1

   iii) 6.25; 46.1 **iv)** $\dfrac{13}{36}$

Paper B page 194

1. 0.8181
2. **i)** median 78, LQ = 70, UQ = 84
   ii)

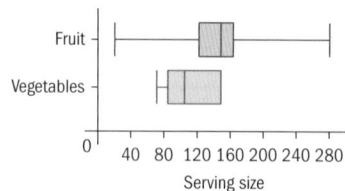

Serving size

3. **i)** $35.8 = \dfrac{(3 \times 5) + (5 \times 15) + 30f + (4 \times 55) + (3 \times 95)}{15 + f}$
   $\Rightarrow f = 10$
   ii) 693.36
4. **a)** 0.544 **b)** 0.0032
5. **i)** 0.04 **ii)** 0.64 **iii)** 0.455
6. **i)** Proof **ii)** $\dfrac{14}{55}$ **iii)** 4.09 **iv)** 0.295
7. **a)** **i)** 3 **ii)** 28 **iii)** 243 **b)** 302 400

Data sets

The following tables give the full data set for the decathlon scores that we have been using in Chapter 3 and that we considered in *Maths in real-life: Sporting statistics*. The first table shows the result that each athlete achieved, and the second table shows the points that they were awarded. Given the full set of data, we can see that there is a surprising amount of variation across events.

You can use these data to create your own diagrams and summary statistics, to explore the data for extra practise at the different techniques and see what other stories the data tell if you were interested in the stories you saw in the Maths in real-life spread and in the exercises in chapter 3.

There are 10 events for each athlete so you can treat each athlete's points score as a set of data with 10 values; there are 75 athletes for each event and you have times or distances for each discipline as well as the points score. Do the 'stories' in the data appear the same when you use the raw event results (times or distances) as when you use the points scores? You could use all 75 athletes to make comparisons or you could consider block – for example, by comparing groups of this ranked list you may get a richer story (you could use 3 blocks of 25 or 5 blocks of 15 to do this).

	Athlete	100 m	Long	Shot	High	400 m	110h	Discus	Pole	Javelin	1500 m
1	Ashton Eaton	10.21	8.23	14.2	2.05	46.7	13.7	42.81	5.3	58.87	254.48
2	Roman Šebrle	10.64	8.11	15.33	2.12	47.79	13.92	47.92	4.8	70.16	261.98
3	Tomaš Dvorak	10.54	7.9	16.78	2.04	48.08	13.73	48.33	4.9	72.32	277.2
4	Dan O'Brien	10.43	8.08	16.69	2.07	48.51	13.98	48.56	5	62.58	282.1
5	Daley Thompson	10.44	8.01	15.72	2.03	46.97	14.33	46.56	5	65.24	275
6	Jürgen Hingsen	10.7	7.76	16.42	2.07	48.05	14.07	49.36	4.9	59.86	259.75
7	Bryan Clay	10.39	7.39	15.17	2.08	48.41	13.75	52.74	5	70.55	290.97
8	Erki Nool	10.6	7.63	14.9	2.03	46.23	14.4	43.4	5.4	67.01	269.58
9	Uwe Freimuth	11.06	7.79	16.3	2.03	48.43	14.66	46.58	5.15	72.42	265.19
10	Trey Hardee	10.45	7.83	15.33	1.99	48.13	13.86	48.08	5.2	68	288.91
11	Tom Pappas	10.78	7.96	16.28	2.17	48.22	14.13	45.84	5.2	60.77	288.12
12	Siegfried Wentz	10.89	7.49	15.35	2.09	47.38	14	46.9	4.8	70.68	264.9
13	Eduard Hämäläinen	10.5	7.26	16.05	2.11	47.63	13.82	49.7	4.9	60.32	275.09
14	Dmitri Karpov	10.5	7.81	15.93	2.09	46.81	13.97	51.65	4.6	55.54	278.11
15	Aleksandr Apaichev	10.96	7.57	16	1.97	48.72	13.93	48	4.9	72.24	266.51
16	Frank Busemann	10.6	8.07	13.6	2.04	48.34	13.47	45.04	4.8	66.86	271.41
17	Dave Johnson	10.96	7.43	14.61	2.04	48.19	14.17	49.88	5.28	66.96	269.38
18	Grigori Degtyaryov	10.87	7.42	16.03	2.1	49.75	14.53	51.2	4.9	67.08	263.09
19	Chris Huffins	10.31	7.76	15.43	2.18	49.02	14.02	53.22	4.6	61.59	299.43
20	Torsten Voss	10.69	7.88	14.98	2.1	47.96	14.13	43.96	5.1	58.02	265.93
21	Michael Schrader	10.73	7.85	14.56	1.99	47.66	14.29	46.44	5	65.67	265.38
22	Guido Kratschmer	10.58	7.8	15.47	2	48.04	13.92	45.52	4.6	66.5	264.15
23	Leonel Suarez	11.07	7.42	14.39	2.09	47.65	14.15	46.07	4.7	77.47	267.29
24	Steve Fritz	10.9	7.77	15.31	2.04	50.13	13.97	49.84	5.1	65.7	278.26
25	Maurice Smith	10.62	7.5	17.32	1.97	47.48	13.91	52.36	4.8	53.61	273.52
26	Bruce Jenner	10.94	7.22	15.35	2.03	47.51	14.84	50.04	4.8	68.52	252.61
27	Robert Zmelik	10.62	8.02	13.93	2.05	48.73	13.84	44.44	4.9	61.26	264.83
28	Michael Smith	11.23	7.72	16.94	1.97	48.69	14.77	52.9	4.9	71.22	281.95
29	Andrei Krauchanka	10.86	7.9	13.89	2.15	47.46	14.05	39.63	5	64.35	269.1
30	Dean Macey	10.72	7.59	15.41	2.15	46.21	14.34	46.96	4.7	54.61	269.05
31	Christian Plaziat	10.72	7.77	14.19	2.1	47.1	13.98	44.36	5	54.72	267.83

	Athlete	100 m	Long	Shot	High	400 m	110h	Discus	Pole	Javelin	1500 m
32	Aleksandr Yurkov	10.69	7.93	15.26	2.03	49.74	14.56	47.85	5.15	58.92	272.49
33	Jon Arnar Magnusson	10.74	7.6	16.03	2.03	47.66	14.24	47.82	5.1	59.77	286.43
34	Lev Lobodin	10.66	7.42	15.67	2.03	48.65	13.97	46.55	5.2	56.55	270.27
35	Sebastian Chmara	10.97	7.56	16.03	2.1	48.27	14.32	44.39	5.2	57.25	269.66
36	Pascal Behrenbruch	10.93	7.15	16.89	1.97	48.54	14.16	48.24	5	67.45	274.02
37	Attila Zsivoczky	10.64	7.24	15.72	2.18	48.13	14.87	45.64	4.65	63.57	263.13
38	Paul Meier	10.57	7.57	15.45	2.15	47.73	14.63	45.72	4.6	61.22	272.05
39	Igor Sobolevski	10.64	7.71	15.93	2.01	48.24	14.82	50.54	4.4	67.4	272.84
40	Siegfried Stark	11.1	7.64	15.81	2.03	49.53	14.86	47.2	5	68.7	267.7
41	Aleksandr Pogorelov	10.95	7.49	16.65	2.08	50.27	14.19	48.46	5.1	63.95	288.7
42	Francisco Javier Benet	10.72	7.45	14.57	1.92	48.1	13.83	46.12	5	65.37	266.81
43	Kristjan Rahnu	10.52	7.58	15.51	1.99	48.6	14.04	50.81	4.95	60.71	292.18
44	Sebastien Levicq	11.05	7.52	14.22	2	50.13	14.48	44.65	5.5	69.01	266.81
45	Yuri Kutsenko	11.07	7.54	15.11	2.13	49.07	14.94	50.38	4.6	61.7	252.68
46	Hans van Alphen	10.96	7.62	15.23	2.06	49.54	14.55	45.45	4.96	64.15	260.87
47	Damian Warner	10.43	7.39	14.23	2.05	48.41	13.96	44.13	4.8	64.67	269.97
48	Valter Külvet	11.05	7.35	15.78	2	48.08	14.55	52.04	4.6	61.72	255.93
49	Eelco Sintnicolaas	10.77	7.27	14.2	2	48.02	14.1	42.81	5.36	63.59	266.98
50	Christian Schenk	11.22	7.63	15.72	2.15	48.78	15.29	46.94	4.8	65.32	264.44
51	Aleksei Sysoyev	10.86	7.01	15.49	2.03	49.1	14.64	54.08	5.1	64.22	278.82
52	Yordani Garcia	10.88	7.36	16.5	2.1	48.77	14.07	43.97	4.8	68.1	286.8
53	Aleksandr Nevski	10.97	7.24	15.04	2.08	48.44	14.67	46.06	4.7	69.56	259.62
54	Jagan Hames	10.77	7.64	14.73	2.19	49.67	14.07	46.4	5	64.67	302.68
55	Konstantin Akhapkin	11.1	7.72	15.25	2.02	49.14	14.38	45.68	4.9	62.42	259.6
56	Stefan Schmid	10.82	7.59	14.14	2.01	48.99	14.2	44.24	5.06	67.63	271.76
57	Olexiy Kasyanov	10.63	7.8	15.72	2.05	47.85	14.44	46.7	4.8	49	264.52
58	Antonio Penalver	10.76	7.19	16.5	2.12	49.5	14.32	47.38	5	59.32	279.94
59	Aleksei Drozdov	10.97	7.25	16.49	2.12	50	14.76	48.62	5	63.51	276.93
60	Kai Kazmirek	10.76	7.45	14.2	2.15	47.04	14.15	43.59	4.96	56.31	273.78
61	Nikolai Avilov	11	7.68	14.36	2.12	48.45	14.31	46.98	4.55	61.66	262.82
62	Mike Maczey	10.99	7.59	14.75	2.06	49.83	14.16	44.56	5.15	62.27	269.93
63	Alain Blondel	11.12	7.5	13.78	1.99	48.91	14.18	45.08	5.4	60.64	260.48
64	Romain Barras	11.09	7.24	15.15	2.04	48.33	14.22	44.51	5.05	65.77	268.43
65	Robert de Wit	11.07	6.98	15.88	2.04	48.8	14.32	46.2	5	63.94	260.98
66	Kevin Mayer	11.23	7.5	13.76	2.05	49.53	14.21	45.37	5.2	66.09	265.04
67	Ramil Ganiyev	10.94	7.58	14.76	2.06	48.34	14.35	46.04	5.3	55.14	276.78
68	Ryszard Malachowski	10.93	7.04	14.94	2.09	47.77	14.34	44.04	4.9	59.58	253.67
69	Deszö Szabo	11.06	7.49	13.65	1.98	47.17	14.67	40.78	5.3	61.94	251.07
70	Sergei Zhelanov	11.04	7.5	14.31	2.13	48.94	14.4	43.44	5	65.9	277.24
71	Dave Steen	11.02	7.56	13.99	1.98	48.22	14.95	44.08	5.2	65.36	261.46
72	Nicklas Wiberg	10.96	7.25	14.99	2.05	48.73	14.75	42.28	4.5	75.02	257.05
73	Ricky Barker	10.72	7.53	14.53	2.1	48.6	14.23	43.78	5.2	58.88	289.56
74	Henrik Dagard	10.82	7.37	14.8	1.97	47.25	14.24	43.24	5	66.34	279.91
75	Aleksandr Grebenyuk	10.7	7.12	15.5	2.02	48.8	14.3	45.52	4.7	71.52	267.3
			7.5704	15.26773	2.058667	48.36853	14.27093	46.75533	4.9596	63.95347	270.8164

	Athlete	100 m	Long	Shot	High	400 m	110h	Discus	Pole	Javelin	1500 m	Total
1	Ashton Eaton	1044	1120	741	850	973	1014	722	1004	721	850	9039
2	Roman Šebrle	942	1089	810	915	919	985	827	849	892	798	9026
3	Tomaš Dvorak	966	1035	899	840	905	1010	836	880	925	698	8994
4	Dan O'Brien	992	1081	894	868	885	977	840	910	777	667	8891
5	Daley Thompson	989	1063	834	831	960	932	799	910	817	712	8847
6	Jürgen Hingsen	929	1000	877	868	907	965	857	880	736	813	8832
7	Bryan Clay	1001	908	800	878	889	1007	928	910	898	613	8832
8	Erki Nool	952	967	784	831	997	924	734	1035	844	747	8815
9	Uwe Freimuth	847	1007	870	831	888	891	799	957	926	776	8792
10	Trey Hardee	987	1017	810	794	903	993	830	972	859	625	8790
11	Tom Pappas	910	1050	869	963	899	958	784	972	749	630	8784
12	Siegfried Wentz	885	932	811	887	939	975	806	849	900	778	8762
13	Eduard Hämäläinen	975	876	854	906	927	998	864	880	743	712	8735
14	Dmitri Karpov	975	1012	847	887	968	978	905	790	671	692	8725
15	Aleksandr Apaichev	870	952	851	776	875	984	829	880	924	768	8709
16	Frank Busemann	952	1079	704	840	893	1044	768	849	842	735	8706
17	Dave Johnson	870	918	766	840	900	953	868	998	843	749	8705
18	Grigori Degtyaryov	890	915	853	896	826	907	895	880	845	791	8698
19	Chris Huffins	1020	1000	816	973	860	972	938	790	762	563	8694
20	Torsten Voss	931	1030	788	896	911	958	745	941	708	772	8680
21	Michael Schrader	922	1022	763	794	926	937	797	910	824	775	8670
22	Guido Kratschmer	956	1010	819	803	907	985	778	790	836	783	8667
23	Leonel Suarez	845	915	752	887	926	955	789	819	1004	762	8654
24	Steve Fritz	883	1002	809	840	809	978	867	941	824	691	8644
25	Maurice Smith	947	935	933	776	934	986	920	849	642	722	8644
26	Bruce Jenner	874	866	811	831	933	869	871	849	867	863	8634
27	Robert Zmelik	947	1066	724	850	874	995	755	880	757	779	8627
28	Michael Smith	810	990	909	776	876	878	931	880	908	668	8626
29	Andrei Krauchanka	892	1035	722	944	935	968	657	910	803	751	8617
30	Dean Macey	924	957	815	944	998	931	807	819	657	751	8603
31	Christian Plaziat	924	1002	740	896	953	977	754	910	659	759	8574
32	Aleksandr Yurkov	931	1043	806	831	827	903	826	957	722	728	8574
33	Jon Arnar Magnusson	919	960	853	831	926	944	825	941	734	640	8573
34	Lev Lobodin	938	915	831	831	878	978	799	972	686	743	8571
35	Sebastian Chmara	867	950	853	896	896	934	754	972	697	747	8566
36	Pascal Behrenbruch	876	850	906	776	883	954	834	910	851	718	8558
37	Attila Zsivoczky	942	871	834	973	903	865	780	804	792	790	8554
38	Paul Meier	959	952	817	944	922	895	782	790	756	731	8548
39	Igor Sobolevski	942	987	847	813	898	871	882	731	850	726	8547
40	Siegfried Stark	838	970	840	831	836	867	812	910	870	760	8534
41	Aleksandr Pogorelov	872	932	891	878	802	950	838	941	797	627	8528
42	Francisco Javier Benet	924	922	763	731	904	997	790	910	819	766	8526
43	Kristjan Rahnu	970	955	821	794	880	969	887	895	749	606	8526

	Athlete	100 m	Long	Shot	High	400 m	110 h	Discus	Pole	Javelin	1500 m	Total
44	Sebastien Levicq	850	940	742	803	809	913	760	1067	874	766	8524
45	Yuri Kutsenko	845	945	796	925	858	857	878	790	763	862	8519
46	Hans van Alphen	870	965	804	859	836	905	776	898	800	806	8519
47	Damian Warner	992	908	742	850	889	980	749	849	808	745	8512
48	Valter Külvet	850	898	838	803	905	905	913	790	764	840	8506
49	Eelco Sintnicolaas	912	878	741	803	908	962	722	1023	792	765	8506
50	Christian Schenk	812	967	834	944	872	815	807	849	818	782	8500
51	Aleksei Sysoyev	892	816	820	831	857	894	956	941	802	688	8497
52	Yordani Garcia	888	900	882	896	872	965	746	849	860	638	8496
53	Aleksandr Nevski	867	871	792	878	888	890	789	819	883	814	8491
54	Jagan Hames	912	970	773	982	830	965	796	910	808	544	8490
55	Konstantin Akhapkin	838	990	805	822	855	926	781	880	774	814	8485
56	Stefan Schmid	901	957	737	813	862	949	751	929	853	733	8485
57	Olexiy Kasyanov	945	1010	834	850	916	918	802	849	574	781	8479
58	Antonio Penalver	915	859	882	915	838	934	816	910	728	681	8478
59	Aleksei Drozdov	867	874	882	915	815	879	842	910	791	700	8475
60	Kai Kazmirek	915	922	741	944	956	955	738	898	682	720	8471
61	Nikolai Avilov	861	980	750	915	887	935	808	775	763	792	8466
62	Mike Maczey	863	957	774	859	822	954	758	957	772	745	8461
63	Alain Blondel	834	935	715	794	866	951	768	1035	747	808	8453
64	Romain Barras	841	871	799	840	893	946	757	926	825	755	8453
65	Robert de Wit	845	809	844	840	871	934	792	910	797	805	8447
66	Kevin Mayer	810	935	714	850	836	948	774	972	830	777	8446
67	Ramil Ganiyev	874	955	775	859	893	930	788	1004	665	701	8444
68	Ryszard Malachowski	876	823	786	887	920	931	747	880	732	855	8437
69	Desző Szabo	847	932	707	785	950	890	680	1004	767	874	8436
70	Sergei Zhelanov	852	935	747	925	864	924	735	910	827	698	8417
71	Dave Steen	856	950	728	785	899	856	748	972	819	802	8415
72	Nicklas Wiberg	870	874	789	850	874	880	711	760	966	832	8406
73	Ricky Barker	924	942	761	896	880	945	742	972	721	621	8404
74	Henrik Dagard	901	903	777	776	946	944	731	910	834	681	8403
75	Aleksandr Grebenyuk	929	842	820	822	871	936	778	819	913	762	8492
		905.1067	952.9467	806.24	858.08	891.84	940.36	803.3067	898.5733	797.84	740.8267	8595.12

Index

Page numbers in *italics* refer to Summary Questions.